The Nature of the Horse

The Nature of the Horse

by Margaret Cabell Self

arco

New York

Published by Arco Publishing Company, Inc.
219 Park Avenue South, New York, N.Y. 10003

Copyright © 1974 by Margaret Cabell Self

Library of Congress Catalog Card Number 73-77840

ISBN 0-668-02993-5

Printed in United States of America

Contents

I

THE EXPERT HORSEMAN

The goal of every serious rider is to become an expert horseman. In this book I shall put before the reader certain knowledge of the mental, temperamental, and physical nature of the horse which every rider who can justifiably be termed an "expert" has had to acquire in order to be worthy of that title and status.

Few people realize exactly what the concept of expert horsemanship implies. So that all my readers may clearly understand the goal toward which they are working, it seems appropriate to begin by delineating the differences between an ordinary rider and an expert horseman.

Riding a horse really well is a far more intricate and delicate art than is generally supposed. A person who can stay on the horse's back at the walk, trot, canter, and gallop is considered by many a proficient horseman, and if he can hop over an obstacle and does not bite the dust when his horse unexpectedly shies, and tries to charge off, he is an expert. To the horseman who knows, however, the first rider described above might be compared to the child who has finished kindergarten, and the second to a pupil who has successfully completed the first grade in grammar school.

Let me explain. Any active young person can learn to sit a horse, either bareback or in a saddle, and to keep his balance by position, relying neither on knee grip nor the support of his reins. He can also learn to maintain this position without interfering with the horse. He will need a good instructor, a suitable horse, and a fair amount of practice, but the problem is much the same as that of learning to ski, ride a bicycle, or perform on the parallel bars, namely, one of balance and coordination. The horse must be a well-trained, docile animal that will take and keep any gait without much help from the rider and not be worried by the ignorance and clumsiness of the tyro on his back. A correct, firm seat is the first basic skill to be acquired, and it must be mastered before the rider can attempt the next step in his education as a horseman—to graduate from being simply a passenger. He must learn to influence and control his horse so that the animal will obey him under all circumstances willingly and effortlessly.

In regard to the second rider we described, whose seat was firm enough to stay on when his horse shied and who had enough control to keep him from taking off, the educated horseman asked to evaluate his knowledge would answer that in such circumstances a truly expert horseman would not have been taken by surprise; he would have anticipated the behavior of the animal and would have prevented any disobedience in the first place. Furthermore, to a bystander he would have appeared to have done nothing at all.

Thus, before reaching "expert" status a rider must be so sensitive to the reactions of his mount that he knows what the horse is about to do a fraction of a second before the horse initiates his action. Also, his own reactions must have become so instinctive that he automatically applies the needed restraining aids (reins, legs, distribution of weight, use of his back, and possibly his voice) to keep the horse moving along smoothly. In other words, he must change the horse's mind for him without inciting him to further resistance.

It is at this stage, when the neophyte has learned to stay aboard a well-trained horse depending only on his balance and his correct position and so is ready to cease being a passenger, that all resemblance to learning to ride and learning to ski, swim, do gymnastics, etc., vanishes. The reason? A new factor has been added, i.e., the need of the rider to completely understand his horse and to learn to anticipate what he is going to do before he does it. Having interpreted the horse's intentions, the rider must successfully transmit his own ideas on the matter, with the result that these ideas are carried out. Riding is the only sport which demands knowledge and skill of this sort.

The horse is a live being, many times heavier and stronger than the rider, with a mind of his own and certain basic behavior patterns that are integral parts of his nature. He also has developed a number of individual responses to various situations, and has a character all his own, distinct from every other horse. All these factors, plus his training, affect his behavior, but regardless of whether the horse is docile, stubborn, or excitable by nature, it is the skill of the rider that largely determines his actual reactions.

Since the horse has a mind of his own, and since the rider with basic skills can become the expert only by acquiring experience, it is inevitable that this future expert, who has passed his first-grade exams with flying colors, will meet with resistance sooner or later—and most probably sooner—which will only be the result of his own inability to understand his horse and communicate with him. The rider should not be discouraged by such events but should realize that with knowledge and practice comes skill. Unfortunately, too many riders have no conception of what lies before them and become frightened or discouraged or, worst of all, put all the blame on the horse when something goes wrong.

Let us take a common example. A rider of the caliber that we have been discussing decides to rent a horse from a local club or stable and go for a quiet hack. He has ridden from this same stable many times before. He knows the territory well. The horse he has rented is one that he has often ridden, and he does not anticipate any problems. However, he has forgotten one thing, namely, that this will be the first time that he has ever ridden out alone. Always before he has either been with a group or with one other companion who was an experienced horseman.

Our rider intends to be out for about two hours and chooses a trail that will carry him several miles from the stable and then, after making a big circle, will bring him back without his having to retrace his steps. There are a number of forks along his way, some of these being points where trails leading back to the stable come in from one side or the other.

The day is fine, the horse seems to move along willingly enough, and the rider relaxes in the saddle letting the reins run through his fingers until they are quite slack and he no longer has any contact with the horse's mouth. They jog along at an even gait. The horse is a "school" horse and does not expect too much from his rider. This is a route which is often taken by the riding instructor and his group of youngsters. Presently less than a mile from home they approach the juncture of a trail which leads back to the stable. It is a turn with which the horse is very familiar, one down which the instructor often turns with his pupils.

The rider sees the turn also but pays no attention since he doesn't intend to go that way. Nor does he notice that the horse has lifted his head, pricked up his ears, and increased the speed of his gait ever so slightly. He is therefore taken completely unaware when, on reaching the fork, the horse darts down the side trail. Before he can gather up his reins his erstwhile docile mount is off at a gallop for the stable.

It may be that our rider being slightly more experienced is able to stop his horse before he actually takes off. Now he may be in a worse predicament. This particular animal has learned the trick of rearing and whirling when he wants to get his own way. In company he never gives trouble for, being a herd animal by nature, he follows willingly enough behind his fellows. Now, however, knowing full well that his rider has little control, he tries his tricks. If, when he goes up, the rider has the presence of mind to lean forward and grab hold of the mane for support without depending on the reins, the worst that will happen will be that the horse will win out and take him back to the barn. Horses that rear from vice seldom actually go over backward, provided their balance is not interfered with. If, however, the rider hangs on to his mount's mouth and lets his weight go back when the horse is at the top of his rear he may disturb the animal's balance causing him to topple over and perhaps land on the unfortunate rider. Should this happen the rider may be seriously hurt and certainly will lay all blame for the incident on the "willfulness and dangerous character" of the horse.

Let us see what happens when the rider is one of experience. A second rider—a horseman—also has hired the horse, one which he has never ridden before, from this same stable. He also knows the territory well and sets out for a two-hour ride. He, like the first rider, is relaxed and has let the reins slide through his fingers but only to the point where the rein is "stretched," not hanging slack, so that the rider is still in communication with his horse.

Long before the horse begins to show any indication of being interested in the side trail, the horseman has instinctively become aware of the possibility of such a reaction. Without conscious thought he shortens his reins and applies any needed aids. As a result the horse passes quietly by the familiar turnoff and a bystander would never have been aware that there had been any question of his not doing so. Who then was to blame when, in the first instance, the horse gave trouble? Not the horse, certainly; he was only obeying his natural inclination to return to his fellows. No, it was the rider and his lack of experience. But if this rider continues his career and makes a real effort to study the nature of his animal, he may in time become as knowledgeable and competent as the second rider—i.e., he may become a horseman.

That these reactions of the rider in such circumstances become truly instinctive I can testify from my own experience. I had ridden for well over ten years before I ever owned or drove a car. For a full year after acquiring one I found myself slightly tightening my grip on the steering wheel and saying "steady now" in a soothing tone to my faithful Model-T each time we approached a familiar turnoff. It could be quite embarrassing when there was another passenger in the car!

How is the rider who has had good basic training ever to reach this status of "expert" horseman? Let us examine what lies ahead. To begin with he must learn, partly from study but mostly from actually working with horses, as much as he can about the physical attributes that affect the way a horse reacts. These include such things as the relationship of his skin to his nervous system, the location of his eyes and how they work, and many other physical characteristics which the rider may never have considered germane. He must understand the horse's basic instincts: These have been stamped genetically and date back to his early ancestor *Eohippus*, the Dawn Horse. One such characteristic, for example, is his fear of stepping on anything which looks odd. There are many, many others. Some the horse has been conditioned to control, but the basic instinct is always there and may take over in emergencies.

The rider who wishes to qualify as an expert horseman must know the horse's mental abilities and limitations. He must understand his various temperamental and emotional characteristics. He must also realize that all of the factors mentioned above must be kept in mind in training the horse as well as in riding him, and he must learn to utilize them for this purpose.

Above all he must keep an open mind, realizing that every horse presents a different problem and that every horse has something to teach him. For myself, after well over fifty years of working with horses, riding them, training them, hunting them, showing them, and handling them in the stable and in the pasture, I am still learning new things with every animal I work with. And I am firmly convinced that there is more to be learned about the horse's nature by caring for him and handling him on foot than by riding him.

It goes without saying, then, that only the rider who has ridden and worked with all types of horses—sensitive horses, phlegmatic horses, unwilling horses, spirited and ambitious horses, highly trained horses, untrained horses, stubborn horses, and spoiled, willful horses—can ever learn how to communicate successfully with and so control any horse of any temperament to the extent that the latter's training will permit.

Generally speaking, during the first half of this century only professional handlers, trainers, and horsemen had the opportunity of working with all sorts of animals. Most riders were content to buy animals suited to their skills and then employ professionals to look after and train them. Of late years, however, many young people have begun getting horses or ponies and taking care of them themselves, often in their own backyards. This is a splendid experience provided that before undertaking such a project they have the opportunity, and take the trouble, to really learn something about the nature of the horse and his needs. Only then will they be able to cope with the various problems which are sure to come up and which have nothing to do with actually riding the horse. These young people must know what to do if they go out to the stable to find their horse down in the stall and unable to get up because he has rolled over too close to the wall. They must know something about the stomach and digestive system of the horse, or they may come into the barn one morning to find that, through ignorance, they have a horse with a serious attack of colic on their hands. They must know how to judge good hay and grain and be aware of the danger of feeding grass which has been recently cut and not yet cured. They must know how to handle a horse when medication is called for and know what care his teeth need. All of this and much more has to be learned.

Pony Clubs and 4H Clubs are doing a great deal toward educating young people in this field; and before parents allow a youngster to undertake the care of a horse or pony they should first make sure that he is qualified to do so.

Although most of the necessary knowledge must be gained by actual experience, a great deal can be learned by study, for theory widens the horizon and gives the pupil a chance to glimpse what lies ahead. It is my purpose in this book to put some of this information before the reader, documenting it as far as possible by experiences I have had myself. I do this with the hope that my readers, being introduced in these pages to some of the many horse characters with which I have had to deal, will learn from them some of the things they must know before they can become expert horsemen.

General Harry D. Chamberlin, author of *Training Hunters, Jumpers and Hacks,* gave what is to my mind a perfect definition when he wrote "An expert horseman is one who can get the *most* out of any given animal with the *least* effort, both on his own and on the horse's part."

There you have it all in a nutshell.

II

THE ORIGIN OF THE HORSE
AND HOW HIS PHYSICAL ASPECTS
DEVELOPED

Eohippus, the Dawn Horse, is considered the ancestor not only of the horse but of various other hooved animals. He roamed the earth 50,000,000 years ago during the Eocene age when the world was very young indeed. He was tiny, the size of a large fox, with four toes on his front feet and three toes behind. These toes were protected by small, individual hooflike nails, but *Eohippus* ran on the soft pads of his feet. He was completely without physical weapons—offensive or defensive—of any kind, since he had no claws, horns, teeth that could rip or tear, or sharp tusks. He could only run from his enemies or hide in the underbrush. His color offered some camouflage, and his senses of smell and particularly hearing and sight were extremely acute, but even with these to help him, at first glance it would appear that this tiny, unaggressive, and delectable morsel stood very little chance of survival. Yet survive he did. He and his descendants have lived on this earth far longer than most of the mammals we know, including man, who of course is a very recent newcomer.

How does any species survive? The only way is by adapting itself to

Eohippus.

and living in harmonious balance with its environment. When the environmental or climatic conditions change, the physical attributes and the behavior patterns of the species must modify as well. Many species have not been able to adapt in this manner and have become or are becoming extinct.

In the insect world, for example, the bee has been one of the most successful species to survive, having developed a society which enables it to live in complete harmony with itself and with its environment. The bee has solved all problems relating to pollution and overpopulation, and has created a balance between supply and demand of its main necessity, food.

The hive is a class society and a mode of living has developed in which its workers, its leisure class, and its reigning Queen accept their roles contentedly and without friction. There seems to be no reason why the bee cannot survive indefinitely, or at least until the atomic bomb or a new glacial age puts an end to life as we now know it. But, in solving its problems, the bee has lost all ambition and, with no challenges to be met, its physical and mental development appears to be permanently arrested.

Now let us see how the descendants of the defenseless little creature *Eohippus* managed to survive for 50,000,000 years. How did they escape destruction during the repeated and terrible ice ages and then survive the melting of the ice caps and the resulting inundation of whole continents? How have such varied climatic conditions as the windswept cold of the Mongolian and Russian steppes and the hot sandy deserts of Africa and Asia affected their physical development? In a later chapter we shall discuss the evolution of the behavior patterns established in the primeval horses and how these affect the behavior of the modern horse. Meanwhile, is the horse of today like the bee? Has his adjustment to a way of life as we know it become so perfect that he will never develop further but remain as he now is in looks, habits, and intelligence? And what exactly were the most important factors which influenced the primeval horse enabling him to adapt to his environment and so survive?

Probably the most important of these characteristics was that he was by nature nomadic and not bound to a specific territory. He needed an ample supply of food to give him the energy necessary for speed in escaping from his enemies. His food for many millions of years was foliage, and later grass, and both of these are bulky yet low in protein. The horse's stomach is very small in proportion to his size so in the wild he must graze almost continuously to provide himself with the quantity of energy-producing protein he needs. As fast as one supply of fodder was used up he wandered on to the next. His lack of weapons also contributed to his nomadic behavior. When the saber-toothed tiger took over a territory, *Eohippus* or his descendants had no choice but to move on.

This behavior pattern continued on through the ages The fossils of *Eohippus* are found only in the western United States and in western Europe. But those of his many and varied descendants have been discovered in every continent of the globe, excepting Australia. Since there was never a land connection between Australia and the other continents it was manifestly impossible for the little creatures to have made their way there.

We shall not discuss each of the descendants of *Eohippus*. Some are the direct ancestors of the horse and others, which did not survive, of course are not. Instead we will skip to one which lived in the Pliocene age a mere 5,000,000 years ago.

His name was *Pliohippus* and he was the first of the primeval horses whose leg bones had begun to resemble those of the modern horse. As the surface of the earth changed and became less boggy, the lateral toes which *Eohippus* sported no longer touched the ground and for lack of use gradually dwindled away. One went all the way up the leg and remains only as a callosity on the inside known as a "chestnut." The other two got shorter and shorter until only narrow bones called "splint" bones remain. The center toe developed a horny hoof and an insensitive sole which was a great improvement as far as protection went because it could be used as a defensive weapon.

Though many fossils of *Pliohippus* have been found in both North and South America, for some reason not yet understood the environment became hostile to him and the species became extinct on those continents. Fortunately plenty of his relatives and descendants survived in Europe and Asia.

How does a given species manage to change its physical aspects sufficiently to adapt to environmental changes? One must first remember how extremely gradual these changes were. It took the descendants of *Eohippus* 35,000,000 years to change to grass-eating animals. This was brought about by the advent of good pastureland in the Miocene age. Grass requires a rotating type of mastication which in turn necessitates a different type of dental structure. Because an early descendant of *Eohippus*, an animal called *Merychippus*, was able to adapt to this change in diet the species survived.

Why and how did he survive? And how was it possible for a whole species to undergo such basic and major changes in its bone and tooth structure?

The answer, of course, is fairly simple. This evolution or ability of a species to modify physical attributes as demanded by environmental changes is the inescapable result of the application of nature's basic law, the survival of the fittest and only of the fittest.

In each successive generation, only those descendants of *Eohippus* which, through slight but important physical modifications, managed to adapt to the gradually changing terrain and to the advent of grass, survived to become the progenitors of the next generation. Those who could run the fastest because their feet were gradually becoming tougher and their useless lateral toes were becoming smaller and smaller survived. And those who could best grind up the new type of food survived. Those who could not were killed off before they mated. Thus the children and the children's children were fathered only by members of the species who had adapted most successfully and inherited their genes (which had mutated in some way) and only theirs. Thus there was no retrogression, for the genes, which govern heredity, which were passed on came only from those animals most advanced in their evolution, never from those which had not succeeded in adapting.

The same principle is used in modern selective breeding programs to improve certain desirable characteristics in domestic animals. Thus, for example, we have widely differing breeds of dairy cattle each catering to a specific demand. The Holsteins, for example, have been bred to produce vast quantities of milk, while the Jerseys and the Guernseys produce far less milk, but produce milk that is extremely rich in butter fat, much more so than that of the Holsteins.

An even more recent and sudden change has taken place in the structural formation of the skull of the collie. Because, a few decades ago, fanciers of these dogs decided that long, narrow skulls were aristocratic and beautiful, today it is virtually impossible to find a registered collie with the old-fashioned broad skull.

And look what has happened to the Shetland pony. First used in the coal mines of England, he then became popular for driving the Governess' cart. Brought to this country, a slighter type Shetland more suitable as a riding pony was produced by introducing Welsh and other blood. Then the breeders got the idea that the American Shetland should become a fine harness pony, high stepping, nervous, and suitable only for showing either in harness or stripped. Today "purebred" Shetlands which can be registered must have at least 50% Hackney blood.

A quartet of prizewinning English Shetlands. These are the descendants of *Equus przewalskii* and were originally bred to pull the coal carts in the mines of Great Britain. They were also used as driving ponies but, because of their thick shoulders and flat withers as well as their short, choppy gaits, they were never popular as riding ponies.

Here is Jessica, the American Shetland intended as a children's mount. She came from the Belle Meade Farm in Virginia. The rider, Gincy Self at the age of two and a half, is perfectly safe on her when she goes out for a ride with Daddy. So trustworthy is Jess that not even a lead rein is necessary. Shoebutton, Jess' mate, also from the Belle Meade Farm, is shown on page 204.

Introduction of Hackney blood has changed the erstwhile child's pony into a miniature fine harness animal. In the change, though, he has become refined and has lost much of his stamina and hardiness. If bred and trained for showing, the high spirits and sensitivity that he has developed have made him completely unsuitable for a child to handle.

Consequently, most are no longer suitable as mounts for youngsters. Yet the American Shetlands of the 1930s and 1940s were ideal for this purpose.

Let us now see how the different types of environment to which the descendants of *Pliohippus* were exposed affected the physical and temperamental aspects of the modern horse and how this knowledge can help us.

After *Pliohippus* came *Equus*, the first truly modern horse, who developed about 1,000,000 years ago. There were four distinct *Equus* types and they developed in four different parts of the globe. From one or the other of these four ancestors come all of the many breeds which are to be found in the world today. Two of these lines developed under similar climatic and environmental conditions and so they and their descendants were much alike. The other two were very different both from each other and from the first two.

Equus przewalskii closely resembled the caveman drawings. He survived up to the present century. He developed in Mongolia and in the steppe country of Central Asia. *Equus tarpanus* which developed in Russia was very similar. Both were short-legged big-bodied rough-coated, and coarse. Both were extremely sturdy and could survive through months of cold in a stony, snow-covered terrain when their only food was bark, small branches, and the roots of trees for which they had to dig through hard-packed snow or ice. The physical attributes which they needed to survive have been passed down to their descendants, and, though sometimes modified by selective breeding, are easily recognized.

These characteristics include thick, heavy hides and heavy coats which protect from the cold. The formation and shape of their skulls and their breathing apparatus is distinctive. The nostrils are mere slits but their nasal passages are extremely large, thus producing what is known as a "ram's" profile. The heads are set on at an acute angle. This physiognomy resulted from the need for protection against the icy air. It would have endangered the lungs if such cold air had been inhaled in

Equus przewalskii.

great volume unless it was slightly warmed in some way. Hence, the narrow nostrils which limit the intake, the large nasal passages which give extensive warm surfaces which effectively raise the temperature of the comparatively small amount of air therein, and the further restriction of the air passage by the narrow, angulated throttle. Since all horses breathe only through their noses, there being no passage of air at all through the mouth, these features are most practical.

Just as the heads of these horses developed characteristics which enabled them to withstand the rigors of the climate, so too did the tail have its own individuality, for it was coarse and heavy and carried tucked in against the buttocks for warmth.

The teeth of these horses and of their modern descendants have extremely long roots and are themselves very hard. The jaw muscles are tremendously powerful, thus enabling their owners to chew roots and saplings thoroughly. Complete mastication is necessary because the digestive juices of the horse are weak, and unless the food is thoroughly ground it passes through the stomach without being digested.

A few years ago an experiment was tried which involved introducing Arabian blood into herds of northern ponies with the idea of refining their appearance. The experiment was a total failure, because though the progeny retained the strong muscles of the jaw and the instinct to masticate powerfully, the teeth lost some of their strength and the roots were shortened. None of the offspring lived beyond the age of eight, having by then ground their teeth away to nothing, and being left with only bare gums, which were useless for chewing even grass or soft grains, they starved to death.

The descendants of *Equus przewalskii* and *Equus tarpanus* include among others the Japanese, Mongolian, and Korean horses, the Celtic horse found in Britain by Caesar, and the Norwegian and Icelandic ponies. It was the Mongolian ponies that the British used in India when they played polo. They were small, 12 to 13 hands, swift, agile, and strong though by heredity not suited to the climate. Today, the Icelandic ponies are even more hardy. After being ridden at a gallop for miles in sub-zero weather they arrive at their destination wet with sweat and smoking hot, to be immediately doused with a pail of water! This freezes instantly into an ice overcoat, and thus protected the animal cools out slowly without foundering! A novel treatment and one which I should hesitate to try on any but an Icelandic pony!

Equus robustus developed in the grassy plains and the forests of Europe. There was plenty of food, good cover, and a moderate climate. He became very large and massive due to the richness and plentifulness of the fodder not only in summer but all year round. Because of his great size there were few carnivorous animals which dared attack him, so speed was not of primary importance nor was extreme alertness to danger. No doubt he would either have become extinct or have developed very differently had he wandered into southern Asia. His immediate descendant was the "Great Horse" of Medieval times. Not only natural selection but the demand for as big and sturdy an animal as possible contributed to his development; for this was the horse that the armored knights used in battle. A knight with all his accoutrements weighed about three hundred pounds, and it took a large, strong, well-balanced horse to carry him. Furthermore, when two opponents competing in a jousting match met head on at the gallop, the heavier the horse the better.

Throughout this era these horses were highly prized, protected, and bred to improve their natural characteristics. From them came the modern draft breeds, which although not valued for their weight-carrying ability particularly, are prized for their strength and docility, for their work is to pull heavy loads easily and plow rough and heavy soil. The draft breeds include the Percheron, the Shire horses, the Belgians, and many others. These descendants of *Robustus* and the descendants of *Przewalskii* and *Tarpanus* are classified as "cold-blooded" horses, being less sensitive and more even-tempered than the "hot-blooded" desert horses.

Of the draft breeds, only the Percheron has a little hot blood in his ancestry. This came about when the Moors overran France after defeating Spain. They brought their desert-bred horses with them and when, at Tours in 732 AD, they were defeated by Charles Martel, they retreated leaving some of their mounts behind them to be crossed with the native horses. Tours is in a corner of a district called "La Perche." The Percheron, a magnificent dappled gray animal, the result of this cross, comes by his name from his place of origin.

The Normans introduced the Great Horse to England and thus the Shires with their tremendously plumed legs came into being.

A six-horse hitch of Clydesdales being exhibited at the Santa Barbara Horse Show. Not as large as the Shires and Percherons, these horses, like them and the other draft breeds, are descended from the Great Horses of the Middle Ages. The latter, in turn, trace back to *Equus robustus*.

The English coach horse derives from the Great Horse of the knights which were introduced into England by the Normans.

The profiles of the draft breeds vary from those of the ponies which originated in cold climates, being straight or only slightly "roman" nosed rather than truly ram-faced. Their muscles are rounded and bulging. Their feet are large, round, and plate-shaped so they don't bog down in wet country.

After the age of the knights, the enormous horses were used only for farming, and other blood from the East was introduced to produce a somewhat lighter horse which became popular for ordinary riding and for pulling coaches and other light vehicles. It was to mares of these modified descendants of *Equus robustus* that the three famous "desert stallions," the *Godolphin Barb*, the *Darley Arabian*, and the *Brierley Turk,* were put, to bring about the origin of the English Thoroughbred. The Thoroughbred was and is valued primarily as a race horse. He inherits his size and strength from his European ancestry and his speed from his desert forebears. Most hot-blooded horses in Europe and the Americas trace their ancestry back to one of the three desert stallions.

We cannot leave the draft horse without mentioning his importance as a modifier of the Thoroughbred blood in the breeding of hunters. The hunter must be active and swift, but he must also be sensible, sturdy, and not too hotheaded. This is particularly true in such hunting countries as parts of Ireland, England, New England, and Canada where the terrain is trappy, sometimes boggy, and often rocky. The horse that can come through successfully must be very sturdy. He must be able to get himself out of trouble when he finds himself in an awkward predicament, by jumping off his hocks from a standstill, which puts a great strain on his tendons and ligaments, especially those of the back legs. Our hunter must not be excitable, but be wise and calm, and he must be able to judge takeoffs and landings and keep his feet out of the way of sharp stones. He must be a truly unusual and versatile animal.

The purebred draft would not be fast enough nor agile enough to make a good hunter. The Thoroughbred is ideal in open country but often too excitable and too fragile to stand up under rough going. But, introduce a little draft blood by putting a Thoroughbred stallion to a draft mare and then breed the progeny to another Thoroughbred, and a good, sturdy middleweight or heavyweight animal is produced, which can keep up to hounds in trappy country and keep himself and his rider out of trouble.

Often a lighter but still sturdy hunter can be produced by crossing once more with the Thoroughbred, and even the progeny of a first cross sometimes produces a hunter that has enough speed to qualify and is valuable for showing in the heavyweight divisions.

And now we come to the last of the progenitors of our horses of today, *Equus agilis*, the horse of the hot, dry plains of Arabia and Africa. Many of his descendants also wandered north into southern Asia where they mated with descendants of *Przewalskii* and *Tarpanus* which had wandered south. But first let us consider how the climate and the terrain affected the physical development of this descendant of *Pliohippus*.

It is only logical to expect *Equus agilis* and his descendants to differ radically from the horses which we have examined up to now. Take the conformation of the head. Since there was no need to warm the already hot desert air before it entered the lungs, the large air chambers in the nasal passages which gave the ram profile to the northern-bred horses were unnecessary, and since there was little cover and many enemies, great speed was essential, which called for a tremendous supply of oxygen, as well as well-developed heart and lungs. So *Equus agilis* and his descendants have wide-flaring and very flexible nostrils which can admit great drafts of air. This is carried through adequate but not enlarged nasal passages so that the head of the desert horse with its wide muzzle, its small nasal passages, and its broad, intelligent forehead has a convex or "dish-faced" profile. Nor is the head "put on like a hammer" with an acute angle at the throttle. Instead the throttle is large and curves gracefully, allowing a better passage of air and permitting the head to move more freely.

Since there was little cover in the desert country for any animal that could not flatten itself and crawl on its belly, *Agilis* had to be ever on the alert. His descendants have the large, bulging eye which gives tremendous peripheral vision as opposed to the much smaller and less rounded eye of those breeds which did not have to spot their enemies at great distances.

Since his domestication the desert horse has been encouraged by his nomad masters to carry his head low as he walks, swinging it from side to side. This was because the rider knew that the horse, with his ultrasensitive hearing and smell, could spot an enemy far sooner than a

Witez, a beautiful Polish Arabian stallion. He is a descendant of *Equus agilis*, the Desert Horse. (Photo courtesy of Mrs. E. E. Hurlburt.)

man could. He watched his mount carefully because at the slightest whisper of sound, perhaps the tap of a distant hoof hitting against a rock, the dainty ears would swing forward and the head would turn toward the direction from which the sound had come. And of course there is no horse in the world that can collect himself and be off as quickly as the Arabian.

Fodder not being very plentiful in this hot, dry climate, *Agilis* and his descendants never grew very large. Fourteen hands to fourteen-two (56 to 58 inches at the withers) is considered an average height and anything over fifteen hands is unusual.

Because of the heat, the skin of the desert horse is thin, and when he has been running the veins stand out like cords on his deep, sloping shoulders. The hairs of his coat and of his mane and tail are short and fine, and he carries his tail away from his body to permit the free circulation of air. These characteristics are designed for cooling and are in direct contrast to those of his northern cousins.

There are rocks in the desert, and the ground can be very hard. Galloping on hard ground puts a far greater strain on tendons, bones, and feet than does running over soft, grassy pastures. So the desert horses' dainty feet have exceptionally tough walls and well-developed frogs. The cannon bones in his lower legs are extremely dense in structure, the cavity running through the center being smaller in diameter and the surrounding bone thicker than are those of other breeds. The tendons are well-defined and stand away from the bones so that the cannons viewed from the front appear narrow and from the side very wide. All this makes for strength and reduces the liability to injuries such as strains, splints, and bowed tendons.

As we have said, there being little cover in the form of forests or shrubs in the desert, the ability to get started quickly and to run far and fast was essential to the survival of these horses, necessitating an ample supply of oxygen. So the desert horse is wide between the forelegs and deep from withers to brisket thus providing enough room for the size of heart and lungs he needs.

There are many different breeds of desert horses and related species, but the Arabian is the most popular, especially for importation to the U.S. and other countries. His history and the records of his bloodlines (which, incidentally, are carried through the female and not the male line) go back to before the days of Christ. The Arabs have used selective breeding to produce a horse that is beautiful, docile, intelligent, and easily trained. They have bred a horse that rarely succumbs to the more common types of injuries to bones or tendons, seldom develops respiratory ailments such as heaves, and above all has tremendous stamina. All this is proved every year in the Vermont Hundred Mile Trail Ride, a competition in which horses of many breeds compete, and which is judged on the condition of the horse after the test. It is almost always won by an Arabian.

As an example of the extraordinary powers of this horse to recover from mistreatment, I was recently told that the winner of this ride in 1972 was an Arabian which had been deprived of both food and water until he was so weak that he could not get up. The matter was reported to the A.S.P.C.A., which sent a dealer to buy the horse for ten dollars.

(Unfortunately, the owners could not be prosecuted because there was plenty of hay and grain on the premises and they declared that the horse just wouldn't eat or drink. Actually, the child whose horse it was had gotten bored with caring for her horse and the parents took the attitude that it would be a good lesson to her if the horse died!) My friend learned that the dealer had the horse and that he was eating and apparently gaining strength. She bought him, nursed him along, and finally got him into such hard condition that he was able to win the ride. Few other breeds, except for some of the pony breeds, would have been able to come back in this fashion. More probably they would have developed blood worms or their resistance to respiratory ailments or to injuries would have become so low that complete recovery would have been impossible.

Outside his native land the Arabian is most prized for his prepotency, his ability to pass his valuable characteristics on to his progeny when crossed with mares of other breeds. For this reason Arabian blood is often introduced to improve other strains.

Since selective breeding is still going on and since showing, racing, and the demand for pleasure animals has brought the population of the horse to its highest point in history, we can safely say that, unlike the bee, the horse has not reached a dead-end in his development, but is still changing to meet the ever higher standards.

III

SYSTEMS OF THE BODY

One of the purposes of this book is to acquaint the reader with the physical aspects of the horse and how his systems function. In this and in the following chapter, we shall take up various illnesses and ailments to which the horse is susceptible. How some of these may be avoided by proper stable and feeding routines will be discussed. Methods of handling the horse in giving medication and what to do before the veterinarian comes are also suggested. Above all we wish to emphasize that some of the most serious afflictions—founder, colic, heaves—can usually be prevented by the watchful and knowledgeable owner. Generally by the time the veterinarian is called the damage has been done. Full understanding of how the systems of the body work can make the difference between owning a horse that has to be put down shortly after reaching maturity and keeping a companion until he has reached an honorable old age.

The Respiratory System

The respiratory system carries oxygen to the lungs, where by a process known as "osmosis" it is exchanged through the walls of the air sacs for the carbon dioxide produced by the body, and is picked up by the blood's red corpuscles and carried to all the tissues of the body—the heart, the brain, the muscles.

Muscles work by contracting and relaxing. They are attached to tendons, which in turn widen and become ligaments. These cover and enclose the joints. When a muscle contracts it shortens itself, tightening the tendon, which in turn acts on the ligament, causing the joint to close its angle. In contracting, the muscle generates friction and heat as well as a poisonous acid (lactic acid, the same as is found in fermented milk) and it is the presence of this acid which causes fatigue. In the instant of relaxation following the contraction, the oxygen brought by the blood partially neutralizes this acid thus relieving the fatigue to some extent. Thus we see the importance of the respiratory system in relation to the stamina and staying power of horses that are, for example, raced, hunted, and jumped. No wonder that the athlete may sometimes feel acute fatigue and breathlessness before he has fully "warmed up" and before his body has adjusted itself to the demands being made on it. After this has been accomplished and the necessary oxygen which partially relieves the fatigue has been provided by the lungs, the athlete finds himself able to continue. He very aptly calls this "getting his second wind."

The respiratory system is subject to several different types of ailments. Two of these affect the throttle, where the head attaches to the neck. One of these throttle ailments is characterized by a whistling sound when breathing, but only when the horse is going fast, and is not serious unless it develops into *roaring,* a deep sound heard when the horse inhales while galloping or trotting fast. The cause of this ailment, which constitutes an unsoundness, is an abnormality in the larynx due to paralysis of the left vocal cord. This in turn reduces the size of the aperture in the throat through which the air passes. As we just learned, limitation of oxygen results in loss of stamina and excessive muscular fatigue. The effort to get enough air through the limited aperture also can affect the heart. Whistling usually appears in horses under six and may or may not develop rapidly until it becomes roaring. About 35 years ago an operation was found to cure the condition, which involves stripping the lining membrane from the pouch which is behind the vocal cord and allowing the cord to adhere to the walls of the larynx, thus leaving an air space permanently open.

Thickened wind is also a condition in which the air space in the larynx is not large enough to permit a free passage of air. It can occur in horses that are overweight, but is most common as a result of any disease in which there is a great deal of nasal mucous discharge. This includes "shipping fever," influenza, bronchitis, strangles, and sinus infections due to inhaling dust or foreign substances, such as the dried seeds of timothy hay. About five years ago a team of veterinary surgeons gave a demonstration of a type of operation somewhat similar to the one described above in which the thickening was reduced. Both of these ailments have been well known and understood for centuries.

In the books of Surtees, whose best known character was the famous "Jorrocks" we find descriptions of "roarers" that had had tracheotomies performed. In this operation a small opening is made into the windpipe (trachea) and a tube is inserted, thus permitting additional air to pass. through. Owners of such horses were advised never to go hunting without carrying a cork of suitable size in their pockets so that if it became necessary to ford a stream they could first plug up the tube preventing the risk of drowning their noble but handicapped steeds through the admission of water directly into the lungs!

The conditions just discussed should not be confused with *high blowing*. This, unlike roaring, is a sound made only on exhalation. It is due to a malformation of the false nostrils, is not harmful in any way, and is very common in highly bred horses.

By far the most common and serious respiratory ailment is emphysema, commonly called *heaves* or broken wind. We have shown how through the process of osmosis oxygen is exchanged for carbon dioxide via the red corpuscles. If the walls of the air sacs are damaged, this exchange becomes very difficult. The horse can draw in his breath but cannot expel it readily.

Common causes of emphysema are a pocket of pus or congestion of blood in the lungs following an infectious disease such as those mentioned in relation to thickened wind, or an allergy to dust or to a specific food such as oats. It should be emphasized that whereas some horses are much more sensitive to dust than others, dusty conditions, whether related to the feed or to the locale in which the horse works, should be avoided as much as possible.

The symptoms of heaves include a dry cough, which becomes noticeably worse when the victim is exposed to dust or hard exercise, and a difficulty in exhaling resulting in a perceptible double exhalation. In cases of long duration this double exhalation causes the development of a muscle behind the flanks of the horse which he uses in breathing out. The depression running along the side of this muscle is known as the "heave line" and is easily seen. Horses with heaves have large bellies and weak flanks.

The immediate treatment, if emphysema seems to be developing, in fact in any respiratory disease where a cough is present, is the elimination of dust as far as possible. Horses should be kept in the pasture rather than in a dusty paddock or in a stable. All hay and grain should be thoroughly wetted down before being offered, and the hay ration should be reduced. If the grass is good it would be best to cut out the hay ration entirely, at least temporarily. If the condition is not far advanced these practices will alleviate the symptoms to a large extent. If the condition is the result of a disease which has left congestion in the lungs, injections of antibiotics will usually take care of this provided they are begun early enough: therefore if a cough persists for more than a day or so after the horse has recovered from shipping fever, bronchitis, influenza, or other respiratory infection, the veterinarian should be called at once.

For the horse that has an allergy to oats or whose lungs have been seriously damaged, a product called "New Hope" which appeared on the market a few years ago, though it cannot repair damage already done, can sometimes prevent further deterioration and give sufficient relief so that the horse can go on working. This is a feed in the form of pellets and consequently is completely free of dust or of the irritating chaff of oats. There are other similar products advertised as being "complete foods," the manufacturers claiming that any horse can subsist and be healthy on their feed without any hay or other bulk at all. I have not found this to be so. Horses, to be comfortable, need bulk in their diet. Those animals which are deprived of it show their discomfort by restlessness, pawing, kicking, and chewing every available surface. A better method, though one which is a little more trouble, is to substitute beet pulp if the horse cannot tolerate hay. The beet pulp, which is often used for dairy cattle, comes dried and must be moistened thoroughly. It then swells and, though it takes a little while for horses to become used to it, when they do they like it very much. I have seen animals so far gone with heaves that they could not be used at all improve enough to return to normal work when put on a diet of New Hope and beet pulp.

I have been told that there is a certain area in western Canada where heaves is unknown, and that horses suffering from it when transported become apparently perfectly well. However, on being returned to their original homes the symptoms reappear. This only proves that any damage done to lungs is irreversible but that dust or chaff greatly irritates the condition and so makes the symptoms more serious.

In a previous chapter we mentioned that a horse cannot breathe through his mouth. There have been cases in which a horse, having suffered a head injury, has sustained a severe nasal hemorrhage. A well-meaning owner has tried to stop the hemorrhage by plugging up both nostrils, a normal procedure if the sufferer were human. But since the horse can breathe only through his nose this treatment has resulted in death by asphyxiation.

The horse's dependence on his nose for breathing can sometimes be used to advantage by the trainer. A badly trained horse will sometimes be very restive when being mounted. This habit should be cured by training, of course, and will be referred to again in a later chapter. But, meanwhile, if the prospective rider is not sufficiently experienced to control the animal for himself, the person holding the horse to be mounted can take advantage of this inability to breathe through the mouth by putting a hand over the nose just above the nostrils and pressing on one side, closing the nasal tube by pushing it against the nasal bone. This will distract the horse as his attention will be held by his trying to get his breath through the remaining nasal tube, and he will generally forget about the rider who is trying to mount and will stand still.

On this same topic I must tell the story of a horse named Flamingo, who came into the stable at the age of eight. He had been a steeplechaser and was a fine, bold jumper, not difficult to handle in the ring or in class. But take him across country or even into the open pasture which adjoined the stable and he would take off for the stable bracing his neck and setting his jaw. The only way to stop him was by using the ''pulley'' rein effect, in which the rider holds one rein very short in one hand and holds onto the top of the neck or to the mane with that hand, while he pulls the other rein up and to the side. This has the effect of tilting the horse's muzzle up and sideways, thus disturbing his balance and causing him to turn or stop short.

The only way to cure a horse that has developed the habit of bolting is by months of work at the slow gaits under experienced hands until he becomes completely and automatically responsive to the aids.

My younger son, Toby, at this time was about fifteen or sixteen years old. Having ridden all his life he was a bold, strong rider who did not know the meaning of the word fear. He was also something of an acrobat on horseback, being our best gymnastic rider. One of his stunts was to vault off a cantering horse landing on the ground on the left side, give one bounce, and vault over the back of that horse and onto the back of another which was being ridden alongside the first horse, landing behind the rider.

Toby was away when Flamingo joined our herd and did not know of his tendency toward bolting. On a Sunday, when there were no riding classes, the horses, including Flamingo, were turned out in the pasture adjoining the stable. Toby, who had gotten home the night before, announced his intention of riding a bit and went off to catch a mount. A few minutes later I looked out to see him jump aboard Flamingo who was grazing with the rest of the herd about four acres away from the stable. The horse was not even wearing a halter and, of course, immediately took off at a dead run for the barn. I was not worried about what might happen to Toby. Too many times I had seen him purposely head a pony for the barn and swing off just as he reached the cement ramp which led to the stalls. I had even seen him grasp the lintel of the door and let the pony go in without him, the pony, by this time, having slowed down to save his legs on the cement. But Flamingo was no pony. He was a galloping horse startled by having a rider suddenly jump on his back and he could easily be running blind in which case he could crash right into the building and possibly kill himself. I need not have worried. Toby also realized the danger and instead of jumping off he braced one hand on the horse's neck, leaned way forward and grasped the nose just above the nostrils. Then, bending the head around he pinched both nasal passages until they were completely closed and the horse, not being able to breathe, had no choice but to stop. I don't, however, suggest this method of stopping a runaway horse as being practical for any horseman less talented than my son Toby.

The Circulatory System

The circulatory system, as we have seen, is linked to the respiratory system in that it is the blood that carries oxygen to, and carbon dioxide from, the tissues of the body. Blood contains two types of cells or corpuscles—white and red. The white corpuseles fight infection and also aid in healing. The red corpuscles carry oxygen and nutriment to the tissues enabling the body to utilize its food intake in growing, building muscles, putting on weight, and in repairing and replacing injured tissues. The blood carries away not only the carbon dioxide produced in the body but wasted and damaged tissues as well.

The circulation of blood is stimulated by massage, so it is easy to see why grooming is so important to the health and condition of the horse. Not only does it improve the coat by stimulating the oil glands which make it shine and by keeping it clean, it actually puts weight on the horse and helps him to recover from injuries and illnesses. The old adage, "a good grooming is worth a feed of oats" is absolutely true. But the grooming must be thorough. A rubber curry comb should be used with a circular motion all over the body of the horse to bring the dandruff and dirt to the surface. This is followed by brushing the hair with a good brush. The groom should apply pressure and push from the shoulder. A final wiping off with a cloth, attention to feet, eyes, and scrotum, brushing the face and legs with a soft brush, and then running the palms all over the body to be sure that there is no dirt remaining nor any hitherto unobserved minor injuries finishes up the process. In England the circulation is often further stimulated by whacking the body all over with a series of light blows from the palms or even with a grooming cloth. Careful grooming such as I have described if given every morning before work begins will do much to condition and fatten a weedy animal and bring bloom to a dull coat. It will also help to keep the skin free-moving.

The amount of blood available to different parts of the body through the large blood vessels varies greatly. The fleshy and muscular parts of the body receive much more blood and consequently heal much faster than do the legs, which are fed by only one artery in each. A cut or scraped knee or an injury to a shin if the skin is broken can take weeks to repair itself, whereas a body injury which looks unhealable will often heal readily.

I am reminded of what happened to a horse named Easy Going. He was a Thoroughbred but not an unduly excitable animal; in fact he could be hunted or hacked with safety by any rider who had learned the basics of good horsemanship.

Before coming into my possession at the age of eight, Easy Going had had the misfortune to suffer an injury to one eye and lose the sight of it. This limited his vision but in most instances in the hunting field he could take the rider through any type of cover and never brush his knee against a leaning tree nor go too close to the end of a panel.

However, he once managed to tangle himself up in the pasture in a discarded axle with its attached wheels which no one knew was there, and received a two-sided triangular cut on the shoulder blade at least ten inches long on each side; also one of the spokes of the wheel came out while he was struggling and penetrated his flanks.

The veterinarian stitched up the flap of hanging skin but did not hold out much hope for the stitches holding. As to the gaping hole in the flanks he said that fortunately this was at the exact point where an operation to relieve the gas caused by colic was sometimes performed and he did not anticipate any trouble if it were kept clean.

This all occurred the night before Easy and I and some of the other horses were leaving for a summer camp where we would stay two months.

The veterinarian was right on both counts. By next morning the flap had torn loose leaving a large raw triangle of flesh ten by ten by about fifteen inches. The hole in the flank already looked better.

For the next two weeks each morning I bathed the area of the shoulder injury with a mild solution of Lysol. Then I applied Balsam of Peru, which is very healing and whose odor repels flies. After this I brought the flap, which was shrinking more and more every day, into place, covered the whole area with gauze on which I had smeared boric acid ointment so that it would not stick and strapped this down with wide bands of adhesive. Since there was necessarily some discharge I kept the surrounding areas of healthy skin and hair covered with vaseline.

At the end of two weeks what was left of the flap had adhered to the wound leaving wide bands of rough, scabby tissue. In another two weeks the scabs had come away, the hair was beginning to grow back, and Easy was back at work in the classes. By the end of the summer there was only the faintest line of a scar to be seen, and when his winter coat came in this was invisible.

The Nervous System

The nervous system of the horse is similar to that of man. The nerve endings which are located all over the body, just under the skin as well as within the body, send messages of perception (hearing, seeing, touch, heat, cold, pain, pleasurable feeling, taste, etc.) back to the brain where they are interpreted. There are several things pertaining to this that are worth keeping in mind. First is the fact that the seat of the nerve endings near the surface of the body is protected by the skin, and the thicker the skin the heavier the protection. Think of a very thin kid glove as opposed to the heavy sole of your shoe. The former represents the skin of the Thoroughbred, the Arabian, or the highly bred American Saddle Horse. The latter is typical of the draft breeds, the mule, and horses which have little Thoroughbred blood in their line or have not inherited a thin skin from the Thoroughbred blood they have. It is obvious that the flick of a switch or the bare touch of a blunt spur would be punishment to the one and hardly felt by the other.

Nerves are subject to injury just as are any other tissues of the body and are much slower to heal, if they heal at all. When nerves which control muscles are injured, constriction of movement or even semiparalysis as well as complete absence of feeling may result. Whether or not conditions such as this will be permanent cannot be foretold; often they are, but sometimes seemingly hopeless nerve injuries heal completely.

I am reminded of a pony which we had many years ago named Lightfoot. She was a young mare, three or four years old, well-schooled, obedient, and with a cooperative disposition. I bred her to one of our pony stallions. A month or so later she got out on the road at night and was struck by a car. There was no laceration but the injury was in the area of the shoulder and the mare could not or would not put her foot to the ground but held it up and hopped around on three legs. For six months she showed no improvement at all and had she not been in foal I should undoubtedly have had her put down. She seemed happy enough, however, and had developed the "three-legged canter" which one sometimes sees circus horses perform. She even learned to hop the low stone walls between adjacent pastures without ever letting the injured foot touch the ground.

Then she suddenly began to show improvement, first extending her leg so that the toe barely touched the ground when she was standing. Next she began to put weight on it, and finally, just before her foal was born, she recovered completely and though she lived to a ripe old age and was used for jumping, hunting, and showing she never went lame again. The nerve had completely recovered.

The Muscular System

Just as a boxer or wrestler has special exercises which he uses to develop and strengthen the muscles he needs in his profession so have horsemen devised special exercises to condition their hunters and race horses and to give muscular strength, balance, and agility to their pleasure, show, and dressage horses. As humans are right- or left-handed, it might be said that horses are "right- or left-legged." Most horses prefer to gallop on one lead rather than the other, turn more readily in one direction than in the other, and have a smoother trot on one diagonal than on the other. A majority of horses are "left-legged," and there are conflicting opinions as to why this should be. One school of thought says that it is related to the way in which the unborn foal lies in the womb. Another school says that it is a result of the fact that from the time he is halter broken, the foal is normally led with the groom or trainer on his left; consequently, he is turned most often to the left, which results in the muscles on the right side of his body becoming more stretched and in his tending to first step out with his left feet. Possibly both these factors influence the muscular development. At any rate it is something of which the trainer is well aware, and in his regular exercises of circling and turning, of practicing gallop departs, in posting to the trot, etc., he is careful to note any lack of willingness in his pupil to take one lead or the other, or any awkwardness of gait and to double the number of times he asks the horse to execute a pattern or exercise in the direction which he obviously does not like.

The young colt being trained on the lunge is taught to canter in both directions. But the mature horse that has been "broken" just until he does not object to a rider and then simply ridden can be very stubborn about taking up a special lead, usually the right. I have known some horses that have been so reluctant, that no matter how tightly they were circled or how they were bent they would still take up the wrong lead. I have found that in an emergency one can resort to the fact that most unschooled or partially schooled horses will change lead over a jump. The horse that refuses to take up the right lead can be sent over a low obstacle placed in the center of the arena not too far from the end wall. He comes in at the canter on his favored left lead but if as he lands he is turned to the right on the track, he will almost invariably change to the right lead, and he can then be kept cantering as long as he doesn't switch back. If he does switch the same exercise is repeated. Of course cantering to the right on the lunge should be used as a daily exercise also until the horse learns to balance himself properly on this lead.

The horse whose trot is uncomfortable when the rider posts on a specific diagonal shows that it has not been ridden properly with the rider consistently posting as much on one diagonal as on the other. This favoring of one pair of legs has upset the horse's balance and has caused one side of his body to develop more than the other, since it is more of a strain on the horse's muscles to support the weight of the rider when the latter is down in the saddle than when he is on his stirrups. Also a different set of muscles comes into play. Consequently, if the trainer of a young horse does not make it a point to post as much on the left as on the right diagonal, the horse will feel smoother under him on one than on the other. Since most trainers working in an arena make it a point to change diagonals each time they change direction and to rise to the trot as the horse moves his outside front leg and inside back leg forward, coming down into the saddle as they are planted, such horses learn to adjust equally well on both diagonals. But uninformed persons do not make a practice of this and those who work their horses only on trails nearly always, often unaware, come to prefer one diagonal over the other, using the preferred one exclusively.

The muscles of different types of horses vary according to their breeding, which in turn relates to their work. Draft horses do not need speed but do need weight-carrying and pulling abilities. Their muscles are bunchy. Race horses that must keep going at high speed for a mile or more have flat, stringy muscles. The Quarter Horse, developed from Thoroughbred stock imported from England in Colonial days and mated to native Chickasaw ponies, also has a round bunchy type of muscle. This, together with the way his center of gravity is placed, allows him to get off to a fast start and to stop and turn quickly. This breed was developed by the southern Cavaliers before the days of race tracks in that part of the country. Yet the young bloods liked the sport so they made "race paths," two parallel tracks running beside each other with space to turn and stop at the end. The tracks were a quarter of a mile in length (hence the name) and the Quarter Horse is said to outrun any other breed for as long as he can hold his breath!

The hind quarters of Quarter Horses are particularly well developed and it is the only breed which, when photographed for advertising purposes, is usually shown from behind. However, I have sometimes wondered if this is not due to a misunderstanding as to where the name came from, so that owners may think that the term Quarter Horse refers to a part of his anatomy and that that is the view which should receive the attention. The Quarter Horse, though he is still raced today, is most popular as a cutting horse and as a general stock horse both for showing and for work. He is now becoming popular in the east also as a trail horse and sometimes as a dressage, hunter, jumper, or equitation mount. There are more Quarter Horses registered than any other breed, including the Thoroughbred, and the Quarter Horse purses are the largest of any. A Quarter Horse–pony cross often makes a most successful child's mount being about 13:2 to 14 hands, intelligent, not nervous, and not stubborn as well as being active, tough, and agile.

We have explained how the muscles function and how they are related to the respiratory system in that they depend on oxygen brought by the blood to relieve the poisons which cause fatigue. Though most of the muscles of the horse have their counterparts in man, there is one which is distinctly different. This is the very useful *cutaneous muscle (panniculus carnosus)* which lies directly beneath the skin to which it is attached over most parts of the horse's body. It allows him to twitch his hide and so dislodge insects. Without it the parts of his body which he cannot reach with his tail, nose, or feet would soon be covered with bloody sores from attacks by the various types of flies whose stings can even penetrate through suede shoeleather. These include the big horseflies which settle and sting until dislodged, green-headed deerflies which do the same, cowflies which are small and brown and fly around the heads and bellies of horses lighting and stinging and lighting and stinging or settle to suck blood, and the little black stable flies which look like house flies but sting viciously, especially on the ears and belly while the horse is in his stall.

Muscles are sensitive and subject to spasmodic stiffness. Any treatment which stimulates the circulation, such as liniments of varying strength, hot and cold wet applications, etc., plus light exercise if the veterinarian advises it, will usually bring relief in a day or so. Because by nature it is well supplied with blood, a muscle that is lacerated will heal much more readily than will a tendon or ligament. If there is injury to a bone or tendon which immobilizes the muscles for a considerable time, they will atrophy just as they do in humans in such diseases as poliomyelitis. But when the injury heals the muscles will sometimes improve with no further treatment, even though they seem to be so seriously wasted away that recovery would appear impossible.

I would like to cite as an example, the case of Bonny, a mare whose name will appear several more times in this book for she was remarkable for many reasons. We got Bonny as a two-year-old. She came from the west in a carload of horses. She was fine-skinned and sensitive, suggesting that she was at least part Thoroughbred and the rest a good strain of Quarter Horse. She was very intelligent and had a strong personality. Her conformation and her way of moving were good and so becoming to her rider that the pair invariably "caught the judge's eye" in equitation classes. She was wonderful in the hunt field and a good child's open jumper, and if she had certain idiosyncrasies (such as a dislike for spotted horses or ponies) she was always forgiven because of her other good qualities.

When Bonny was twelve years old we decided that these good qualities should be perpetuated and we bred her to our little Thoroughbred stallion, Meadow Whisk (by Broom Whisk out of a great-granddaughter of Man of War).

When her foal Meadow Lark was born he turned out to be a lovely chestnut colt, taking after the Thoroughbred in conformation. He, too, will be mentioned several times later but for the moment suffice it to say that I have never seen another horse that approached him in his disposition and ability to learn. (At the age of 26 he is still in use.) As in the summer when Easy Going was hurt, I again went away for several months. On my return I found that Bonny had been kicked in the pasture and had seriously injured her near hind leg in the area of the fetlock joint; either a bone had been broken or, more likely, a tendon had been torn. Bonny was on three legs, only resting the tip of the toe on the ground. Before the injury occurred she had been bred back and she was still nursing Lark, so it was out of the question to put her down.

Time went on; Bonny still refused to put weight on her bad foot and at the end of about a year from the time of the injury the whole left haunch was completely wasted away, while the right haunch, in spite of a lack of exercise, was still round. Seeing the mare from two different sides was like looking at two different animals.

Just before Bonny's second foal was due she started hobbling, lightly touching the toe of the injured foot to the ground and allowing it to bear a very little weight. Six months later she was walking fairly freely. As the injured leg had definitely shortened, I had a special shoe with a very high heel made for that foot. Sometime during the second year Bonny started trotting and cantering in the pasture and, believe it or not, at the end of the third year her haunches matched, the muscles in the left side having regained their original strength and shape. She no longer limped, though she still had to wear her special shoe, and she was back at work again!

Ligaments and Tendons

Ligaments and tendons are more sensitive to injury than any part of the horse's body except his skin. Although they often completely recover from slight sprains and strains (though it takes time), a veterinarian should be called at once whenever there is severe lameness which appears to result from injuries involving tendons, ligaments, or bones. Treatment will depend on how the trouble is diagnosed and only the veterinarian is qualified to recommend it.

Though such treatment varies, generally speaking, almost any medication prescribed is designed to help nature repair the injury by increasing the flow of blood to the affected part and so promote healing. Its value lies in following exactly the instructions of the veterinarian and in continuing the treatment methodically until the injury is healed or all hope for recovery or improvement is gone. The most common injury to the tendons is one known as "bowed tendons," or tendonitis. It is most prevalent in one or more of the tendons of the forelegs behind the cannon bones. The most common cause is too hard work especially at a young age, and for this reason it is very common in racehorses that have trained for the track and raced as two-year-olds.

If it is treated by a veterinarian early enough, it is sometimes possible to cure acute tendonitis completely. Such treatment will involve injections and keeping the legs in plastercasts followed by at least a year of complete rest. Chronic tendonitis is rarely entirely curable. The classic treatment is pinfiring followed by rest, but many veterinarians feel that it is the rest and not the pinfiring that helps. Horses that have this ailment can often be used for slow work but as soon as they are jumped or given fast work they go lame again. Shoeing with thick rubber pads with an elevated heel sometimes helps.

Bowed tendons can also be caused by bad shoeing. The diagram shows a horse with bowed tendons, which are also recognizable by the distinctive lifting of the heel off the ground because of the shortening of the tendon and pain, soreness, and heat in the afflicted part, and the

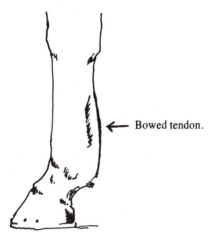

← Bowed tendon.

extreme lameness of the horse. If colts were not put to work too soon this condition would be much rarer than it now is.

The Digestive System

Knowledge of how the digestive system of the horse works is of extreme importance to the horseman since it greatly affects how, what, and when he is fed.

The natural food of the horse is grass. Fresh grass is acknowledged to be highest in vitamin and mineral content of any natural food. However, the mineral content depends on how much the soil in which the grass is grown is depleted. The proportion of protein, or energy and muscle-building content in grass, is low in relation to its bulk as compared with that of grain. In primeval days, after the early horses had become grazing rather than browsing animals, the season when the grass was in seed was the only time when they were able to eat that seed or grain, which was higher in protein. And this season was very short, so most of the year the horse had to depend entirely on grass. This meant that he had to graze almost continuously in order to supply his body with what it needed, and nature designed a system to fit the need. Rather than being held in the stomach or intestines for a long time, food moves through quickly and the horse continues to digest and evacuate as it grazes. A large stomach would be a handicap to an animal that depends on flight as its only defense, so a horse's stomach is very small in relation to his size. As fast as the food in the stomach is digested it passes into the intestines, where the blood picks up the necessary nutriments, and the wastes continue on to the bowels, where they are evacuated as they accumulate. The stomach, meanwhile, is busy on another load of grass.

We must also remember that grass, being soft, is easily digested once it is thoroughly masticated, and so the stomach juices of the horse did not need to be and are not as strong as those of the dog, for example. The horse cannot digest food that is not thoroughly chewed; it simply passes through his stomach and he gets no good from it. The dog, and other carnivores had to eat their kill as fast as they could so that larger animals would not steal it from them. They gulp it down, their teeth being designed for tearing but not for grinding. So nature gave the dog digestive juices strong enough to dissolve even bones when swallowed.

Having developed a digestive system completely adapted to its natural needs, the horse of today now finds himself having to adapt to a different feeding pattern altogether. Instead of grazing on grass all day long, he is fed grain which is hard in texture with a high concentration of protein. Neither it, nor the hay which supplies the bulk he needs is as high in vitamin and mineral content as is natural grass of good quality. The hay also takes more mastication and, if cut too late or kept too long, can be dusty. Timothy if it is cut after it has gone to seed is especially irritating to the horse since the dried seeds can be inhaled and cause either sinusitus or even heaves (see respiratory system). Cut when the seeds are still green it is excellent.

Now let us consider water intake. Water, when it enters the stomach, passes directly through it whether there is anything else in the stomach at the time or not. If there is only hay in the stomach not much harm is done, but if there is grain there and a great deal of water is drunk all at once, it may carry some of the undigested grain with it into the intestines. If not passed on through, this grain may settle in a pocket in the intestines where it ferments. A gas is generated and pain ensues. This is one version of the ailment, common also in humans, known as colic.

Now knowing how the digestive system of the horse works and how he handles his food and water, let us consider how this affects the way we feed him.

First, since we know the stomach is small, we should never feed too much oats at a time. A horse that is not working for more than two or three hours a day generally does very well on two feedings of from two to four quarts each plus about twenty pounds of hay also given in two feedings, the larger quantity being given at night. A horse that is thin or is doing hard work will need three feedings a day at least.

Proper mastication is important, especially of oats. Greedy horses sometimes eat too fast. Therefore the hay should be put before the horse first and he should be allowed to begin munching on it before his grain is given. If several horses are to be fed, giving hay to all and then giving the grain in the same order will allow those hayed first sufficient time to eat enough to take the edge off their appetites before they get their grain ration.

Some horses will eat too fast even though they are not particularly hungry. These gobblers can easily be identified by the fact that they dive into their mangers with their noses scattering grain to the winds. The old-fashioned remedy used to be to put several large stones into the manger and in pushing these aside to get at the grain the horse is constrained to eat more slowly and to masticate better. A similar and double-value method is to put three or four large lumps of rock salt into the manger. He may try and chew these up at first especially if he has not been getting enough salt in his diet, but when the appetite is satisfied he will be contented with licking or munching the salt only occasionally. A special type of manger is also made for horses of this character. It has a series of little pockets on the bottom instead of being simply rounded, smooth iron and the horse must get his grain out in small mouthfuls.

There are also one or two other things to be considered in giving hay. The natural way for a horse to eat is with his head lowered. This also helps drain the sinuses if there is any discharge from the nostrils due to irritation or to infection. However, if the horse is confined in a stall and the hay is just thrown in loose he may not eat it all up at once and may scatter some around as he moves about. This will then become fouled with excreta and urine, and when eaten later may cause trouble. To avoid this many people use either hayracks or haybags. The horse is forced to pull the hay out in small bites which he eats as he goes along. The objection sometimes raised to these is that if the rack or bag is hung higher than the horse's head the dust or seeds may get into his eyes and cause inflammations such as conjunctivitis. Yet if it is hung too low the horse may get his feet entangled in it.

My own experience has been that a haybag or rack can be hung so that the highest part is on a level with or slightly higher than the horse's head and the lower part will then be sufficiently high off the ground for safety. I have never, in fifty years, had a case of conjunctivitis nor had a horse catch his feet in a bag or rack as a result. In feeding hay in paddocks I often use haybags, hanging them from posts. If, however, I have an animal recovering from sinusitis or from shipping fever and there is a heavy nasal discharge I put the hay on the ground to help induce free drainage of the nasal passages. When the horse has finished eating I take away any hay that may have been fouled or trampled, collecting any good hay that is scattered and piling it together for him to come back to.

In stables where many horses are kept and there is not room for the animals to be in box stalls, the matter is very simply solved, for in this case the horses will be in standing stalls. A good standing stall is five feet wide and at least ten feet long. All the way across the front end is built a wooden manger which reaches to the floor. The height should be about that of the point of the horse's shoulder and the manger is narrower at the bottom than at the top so that the horse can stand close to it and reach well into it without difficulty. (Fourteen inches wide at the bottom and about two feet at the top is usual.) Either an automatic water fountain or a pail should be in one corner. In the center there should be an iron or hard plastic grain manger and the hay is dropped in the other corner. Thus the horse cannot scatter his hay, can eat with his head down, and does not need to finish it all up at once.

The exact amount of hay and grain needed by a particular animal depends on several factors, his size, conformation, breeding, temperament, individual metabolism, age, and the work he is doing. A draft animal, even though much larger than a riding horse, will not need as much grain because his work does not require as much energy and because he is not of a nervous temperament. A Thoroughbred, being rangy in conformation and of a very nervous temperament, will always require more grain than will a Morgan or a Quarter Horse. A pony should not be given hard grain such as corn, oats, or "sweet feed" unless he (or she) is either out of condition, in foal, still growing, or doing hard, regular work. Under other conditions grain affects most ponies the way alcohol affects some people. It goes to their heads, they become highly excitable and often very aggressive. One of the feeds consisting of pellets which are mostly compressed alfalfa can be given to ponies that are doing moderate work and need something other than hay but which become too high-spirited for their young riders when grained. A bran mash made of steamed bran, salt, and a little linseed is beneficial after a hard day's work and before a day of rest. But most ponies get along very well on grass in summer and hay in winter. If the pony is sluggish or thin two to four quarts of one of the commercial feeds will provide energy and get him back into condition. These feeds are composed of a variety of grains plus some alfalfa as well as vitamins and minerals. Those which have molasses added for palatability are known as "sweet feeds," and are also very good for horses that are "fussy" eaters. However, in hot weather the molasses tends to ferment if kept very long so unless there are enough horses to use up a bag in a few days it's better to use a commercial feed which is not sweetened during the hot season.

Mares in foal, of course, and growing colts need more grain than do others, and vitamins and minerals are especially important to them. Most stables if they do not use feeds already containing these usually buy them separately and add them to the daily ration of grain.

The type of grain fed may vary with the part of the country in which the horse lives. In the the south and parts of the west corn is often used. In other areas oats are preferred. Many owners prefer to use crimped or crushed oats rather than whole oats since they are easier to digest because they need less chewing.

Hay provides the bulk of the horse's food unless he is out at pasture. It is not as high in protein as grain but, the moisture having been evaporated, it is a more highly concentrated food than grass. Consequently,the horse does not have to eat as much hay as he would grass to get the same amount of nourishment. Too much hay (or grass either for that matter) will result in a "hay belly" and in lack of muscle tone. This is common in horses turned out to pasture.

Good hay is green or pale greenish brown in color. It has a delicious odor, is crisp and not limp in texture, free of dust, weeds or mold, and should contain some of the following preferred grasses, preferably a mixture of two or more: timothy, clover, orchard grass, rye grass, or alfalfa. Of these either timothy and clover or timothy and orchard grass are preferred in the east. Alfalfa is very popular in the west and in Mexico but it is exceedingly rich and can disagree with horses not accustomed to it. If there is any reason why an owner can buy only alfalfa for horses accustomed to other types of hay he should get oat straw and mix the two starting with only a very small percentage of the alfalfa. This situation occurred in New England during World War II. For several months it suddenly became impossible to buy any hay excepting green alfalfa. I was careful to mix mine with straw and to feed it very sparingly but my veterinarian told me that many unknowing owners fed the straight alfalfa with the result that there were a great many cases of diarrhea and scours and some serious incidents of colic.

Hay must be cut while the seeds are still green otherwise much of the nutriment is lost and the hay is almost sure to be dusty. It must come from land that has been kept fertilized or, like grass, it loses much of its food value. A word of caution regarding feeding hay or grass which has been cut and then allowed to stand a few hours but is not yet cured. This often results in colic or even death, for some grasses that are freshly cut develop a toxic substance which vanishes after the grass is completely dried. So when you mow the lawn, feed the grass immediately or else wait until it is dry, brittle, and snaps when you break it before giving it to your horse.

Nearly everyone is aware of the danger of putting hay which is not completely dried out into an airless hay loft. Many fires have been caused by the spontaneous combustion which can be the result. Sometimes one buys baled hay and is not aware of the fact that it has been baled green. A good idea when purchasing hay or on receiving a load is to open a bale or so and feel inside. Long before spontaneous combustion occurs the hay will be hot and this is a sure warning.

How and when water is offered is also important. There are several schools of thought. We have already noted that when the horse drinks, the water passes directly through the stomach, sometimes carrying whatever is in the stomach along with it and that this can cause trouble. Therefore *if the horse is very thirsty*, having been deprived of water over a period of hours or if he has just been working hard but is not overheated, the water should be offered before the hay or grain, and he should be allowed to drink all he wants. If he is very hot he can be given a little water but then should be walked cool before he is put back in his stall and allowed to drink more.

It used to be the custom to give a horse water only in the morning, at noon, and at night, taking the pail away when he was finished. In the opinion of this author and of many veterinarians and owners of horses, a better method is to keep water in the stall at all times and to have it available in an outside trough for horses that are kept at pasture or in a paddock when not being worked. In this way the horse sips water whenever he needs it and never becomes so thirsty that he drinks large amounts at once. When a horse has been working several hours and has then been cooled out he may be put into his stall where his water is waiting, and the hay and grain are then pushed down to him. He should never enter a stall after work to find the grain and hay already there, but should have a chance to quench his thirst first. I have had several horses that actually liked to "dunk" their hay while eating. They would take a mouthful and before swallowing would sip a little water. I have always considered this a good practice since it automatically dampens the hay thus inhibiting the ingestion of any dry dust. I have also noticed that none of these horses that "dunked" ever developed the dreaded disease of "heaves."

There are two methods of keeping water before the horse at all times. One is to hang the water pail three feet off the floor in one corner of the stall and see that it is always kept full. A second, and better method, is to provide automatic water fountains such as are used for cows. Most horses learn to use these very readily and one can then be sure that they always have water when they want it.

Salt as we have mentioned is available either in "rock" form or loose to be mixed with the feed. However, a more common form is the salt brick or block which is put in a special holder on the wall. These bricks can be obtained either as pure or iodized salt or may contain other minerals.

One last word on feeding before going to various ailments of the digestive system. If a horse is to be idle for a day or more always cut his grain ration in half. Failure to do so may result in a serious disease known as azoturia, sometimes called Monday morning sickness, which will be discussed next.

Diseases of the Digestive System

Azoturia and Forage Poisoning. Although the exact cause of azoturia is not known, it is associated with the combination of forced confinement plus too heavy feeding of grain and affects only horses that are in hard condition and accustomed to plenty of exercise. Its symptoms are a rolling motion when the horse first comes out of the stable followed by stiffness in the hindquarters and a tendency to drag the back legs. The horse may be unable to stand and, on falling, struggles violently. His temperature can go up to 103 degrees, his rate of respiration will increase, and his urine will be coffee-colored. Because of the last the disease is also know as Black Water Fever.

This disease is frequently fatal and a veterinarian should be sent for at once. Meanwhile do not move the patient more than is absolutely necessary, blanket him well, and apply heat over the loins and kidneys. Prevention is far better than cure and like founder and thrush this ailment can usually be prevented by good stable management.

An ailment somewhat similar in symptoms to azoturia is fodder or forage poisoning. This may be the result of eating some poisonous plant either while out at pasture or one which has been introduced in the hay. Or it may be a result of being fed grass which has been cut long enough to develop the previously mentioned toxic substances and not long enough to have been cured properly. It may also be due to certain fungi or molds found in hay or grasses. Most horses that are native to a given territory know enough not to eat local weeds or other plants that are toxic, but horses recently imported from a different part of the country may not be familar with the local plants. The symptoms will vary according to the poison, but often start with lethargy and stiffness followed by weakness in the hindquarters. The horse cannot stand and struggles to rise. Mild cases often recover but if the dose is too big or the plant too toxic, death results in a matter of hours. Needless to say, the veterinarian should be called at once.

Colic is the most common ailment of the digestive system and like colic in man varies in the intensity of the pain and in the causes. Colic can be the result of overfeeding, feeding unaccustomed foods without first conditioning the horse to accept them, moldy or bad hay, green oats, parasites, internal injuries, such as a twisted gut or a stoppage in the S bend between the stomach and the small intestine, allowing a horse to drink a great deal of water after having just finished his grain, insufficient mastication of the grain owing to a sore mouth or lacerated jaw from sharp teeth, stones or sand in the bowels, kidneys, or bladder, etc.

The symptoms of colic are unmistakable. The horse becomes very restless, biting at his flanks, and repeatedly lying down, attempting to roll, and then getting up again. Many authorities think that he should not be allowed to lie down but should be kept moving for they think that rolling violently may cause a twisted gut. Others say that such behavior is symptomatic of the horse already suffering from a twisted gut. The horse's belly is generally very distended and there is every indication of intense pain. The cause of this distension and of the pain is the gas which is generated in the stomach or in the intestines because of one of the above irritations. It is the possibility of this gas pressing on the heart which makes the disease so dangerous and it calls for immediate medication if the ailment is not to prove fatal. Since time is a factor, all stable First-Aid kits should contain a bottle of colic remedy prescribed by the veterinarian and a supply of capsules in which the medicine is given.

Giving a capsule is not difficult but an exact technique must be observed. The horse's tongue is first pulled out of the side of his mouth and held there so that he cannot close his teeth. This can be done by an assistant who stands on the off side of the horse. The person giving the medicine and standing on the near side of the horse at his head holds the capsule in his left hand, his thumb curled under to hold it in place against his palm. With the knuckles up and the palm down he now introduces his open hand into the horse's mouth using his fingers to push the capsule well back into the horse's throat. The hand is then withdrawn, the tongue is released, the muzzle of the horse is tilted upward and the jaws held shut. If the gullet is stroked downward hard, it will help induce the horse to swallow and the path of the capsule can be observed as it goes down the gullet. It is of utmost importance that this is done, that there be no mistake that the horse has downed it, for some horses will hold the capsule in their mouths and drop it out later. Nor should the animal be permitted to crunch the capsule before swallowing as many colic remedies contain turpentine or other highly irritating substances which will burn the horse's tongue.

If there is no colic medicine on hand, whiskey, brandy, or gin will often relieve the pain and the gas and an enema of warm soapsuds will help the horse expel gas through the rectum.

Some years ago during the prohibition era when liquor was hard to come by, I heard a horse in trouble in the stable in the middle of the night. I went out to find one of the young horses flat on his back, his belly terribly distended and all four feet sticking straight up in the air and as stiff as pokers. I rushed back to the house, got a pint of gin (highly prized, of course) came back, and poured the whole thing down the throat of the supine animal. Without waiting to see the effect I returned to the house to prepare an enema. I was somewhat surprised, when I came back, to find the patient on his feet and munching hay!

Diarrhea is usually an attempt by nature to get rid of something which the horse has eaten which is irritating to the intestines. Horses put out to pasture early in the spring and allowed to graze their fill on the new grass to which they are not accustomed often develop both diarrhea and colic. Diarrhea may also be due to parasites or be caused by the horse having experienced a chill after being overheated. If the latter is suspected the victim should be well-blanketed and then fed a hot bran mash to which linseed has been added. If unwise diet is suspected give a pint of linseed oil and if the condition persists ask the veterinarian to recommend further medication. *Scours*, a form of diarrhea, is common in foals and is laid to a variety of causes. The veterinarian should be consulted.

Since diarrhea is frequently followed by a few days of constipation a small handful of epsom salts given in the grain every day until the bowels appear normal is a wise precaution.

Parasites are very common in horses. Many veterinarians recommend that all horses be wormed at regular intervals whether the presence of worms is suspected or not. This worming is usually done in the fall after the ground has frozen and in the early spring before it has thawed out as there is less likelihood of reinfection. Since the amount of the dose will depend on the size of the animal and since the method must be followed exactly, the veterinarian should check the animal and then advise accordingly, rather than have the owner buy any vermifuge which may be recommended as ''safe and thorough.''

There are three common types of worms which attack horses. *Intestinal* or *round worms* live in the intestines. They are easily identified when expelled being white, stiff, up to a foot in length and about the thickness of a pencil. In large quantities they cause irregularity in the bowels, loss of condition, and continued loss of weight even though the horse is being amply fed. Another symptom is dullness of coat and a tendency to be ''hidebound.'' They are one of the common causes of colic.

Whip Worms are usually under two inches in length and are very thin. They occur in the rectum. They cause itching of the anus, and the horse will rub his tail against a post at every opportunity sometimes until all the hair on the dock is rubbed off. Examination of the anus will disclose a whitish pasty discharge. A handful of salt in a gallon of warm water given as an enema will usually clear them out. They are the least serious of the worms which attack the horse.

Blood Worms are the most harmful. They actually live in and on the blood thus depriving it of nutriment. Starting in the colon they often pass through the wall into the blood stream where they are carried all over the body. When they become too numerous they may even cause a blockage in the blood stream and hence death. The symptoms are diarrhea with an extremely offensive odor, hollow flanks, dropped stomach, extreme emaciation, anemia, a dry coat, and occasionally lameness. If too far advanced the condition is incurable. Regular worming will prevent this from ever happening.

Bots are a slightly different type of parasite which use the horse as a host. The eggs, tiny, seedlike, and yellow, are laid by the parent fly on the forelegs and sometimes the shoulders of the horse while he is grazing. The bot fly looks like a very small bee with a curled under tail and can often be seen buzzing around the lower front legs of the horse. In licking and scratching his legs the horse ingests the eggs which then hatch. They are not as serious as the blood worms but if there are too many they will take nutriment from the horse and he will lose condition. The veterinarian will recommend a vermifuge that will take care of all the above-mentioned worms at the same time. However, bots are easily prevented by scraping the eggs off the legs as soon as they are noticed. A regular razor blade drawn downward with the hair is best. Horses with long coats should be clipped and the hair burned.

IV

SPECIFIC PHYSICAL CHARACTERISTICS

*How Various Parts of the Body Function; Causes of
Malfunction; Care and Practical First Aid.*

The Head and Neck

The head and neck of the horse assist him to balance himself. The head
must be in proportion to the neck. A heavy head at the end of a long,
willowy neck will make the horse overbalanced on the forehand, just as a
weight attached to the end of a stick will make carrying or swinging it
difficult. A more detailed discussion of how the horse balances and
distributes the seat of gravity will come later. At the moment, the point to
be emphasized is the fact that in traveling over rough ground and in
jumping, the horse should be allowed as much freedom of the head and
neck as possible. Thus, in going up or down steep inclines, the rider
should put his weight on his stirrups and allow the horse to have a free
rein. Also, if the horse stumbles, the rider should allow the reins to slip
through his fingers so that the horse can get his balance and recover
himself.

In an earlier chapter we discussed various types of heads characteristic of different breeds and how and why these developed in different parts of the world. Now let us look at some of the important parts of the head, such as the mouth, the teeth, the eyes and ears and how they function, and then see how this knowledge will help us to understand and to work with our horse.

The Mouth. An examination of the mouth will show that the teeth are arranged in two groups. The grinders or molars are at the back of the jaw; the incisors in the front. Between the two is an area of gum known as the ''bars.'' It is on this toothless area that the bit rests. Were it not for the bars we should not be able to control the horse with the bridle as we know it because the bit would rest on the teeth, which are insensitive. Instead we should have to ride him always in a hackamore, which works on the nasal bones and on the under part of the jaw at the chin groove.

The bars are very sensitive and when the horse is first introduced to the bit every precaution should be taken to see that they are not bruised by

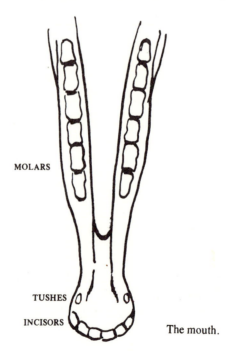

MOLARS

TUSHES

INCISORS

The mouth.

unaccustomed pressure nor calloused by too steady and hard pressure. Bruised bars are intensely painful to the horse and in trying to escape the bit he may become a head thrower or a star gazer with a ewe neck. If the bars become calloused by steady pulling the result may be a horse heavy in the forehand that "bores" on the bit, bringing his chin in to his chest. Should he have a thick neck and be a generally insensitive animal he may take to bolting.

To avoid these conditions it is suggested that the horse spend many weeks being lunged (sometimes spelled longed) at all gaits and in both directions. At first he wears only a lunging cavesson and either a regular bitting harness or a roller or surcingle to which straps with rings have been attached. The reins of the cavesson are fastened to these rings but left so long that they exert no pressure on the horse's nose. In this way he learns the voice commands to walk, trot, stop, turn, and stand. When he has learned these lessons the side reins are shortened just enough so that if the horse throws his head there is pressure on his nose. After a few

A lunging cavesson.

attempts he learns that as soon as he brings his head to a natural position the pressure is relieved.

The next step is to replace the lunging cavesson with a training hackamore. This is not like the ordinary western hackamore which can be severe. It consists of a fleece-lined noseband and a covered curb chain. These are attached to the loops of specially designed shanks similar to the shanks of a curb bit but made of aluminum, flat, slightly curved, and only about three inches long. An ordinary bridle headstall is used with reins fitted as usual to the lower rings of the shanks. The horse is now worked on the lunge in this with the reins being attached to the bitting harness. At about the same time the bit is introduced for the first time. This is done in the stall.

The bit used can be a special type of jointed bit with several "keys" attached or it can be an ordinary snaffle. If a snaffle, it should be fitted so that it hangs a little lower in the horse's mouth than it normally would, but not so low that the horse can get his tongue over it.

The purpose of the keys, or if the snaffle is used of fitting it this way, is to induce the horse to mouth and play with the bit. He will probably do this for several days. To make it even more palatable, it may be rubbed with molasses or syrup before being offered each morning.

When the horse has stopped playing with the bit, the next step is to cross-tie him in his stall or in an aisle, the shanks or tie chains being fastened to the bit, which has now been raised to the proper place in his mouth. The cross-ties should not be tight and someone should stay around the first time to be sure that the horse, in throwing his head or trying to push forward or back up, does not get frightened when he feels the restraint of the bit on his bars.

Then comes the day when the bridle is put on under the hackamore, its reins being fastened to the bitting rig and the lunge line still being attached to the hackamore. The colt now has his daily lesson and with this arrangement, it is impressed upon him that as long as he does not throw his head or try to escape the light touch on his bars, he will not be made uncomfortable but that if he does, fighting does no good and only by relaxing will the uncomfortable pressure be relieved.

When the time comes to mount the young horse for the first time, the above rig, snaffle under hackamore, is left on. The trainer uses all four reins but leaves those attached to the snaffle longer than those attached to the hackamore and lets them come into his hands between the ring and middle fingers with the hackamore reins on the outside.

Not until the colt has learned, under saddle, to move out freely at the command of the legs and to decrease his gaits willingly should the rider attempt to use the snaffle to control his horse. Even when he has become responsive to the bit, when he is introduced to jumping or when he is ridden by people whose hands are not of the best, he should be ridden in a hackamore. I have had horses trained as suggested above that at the age of twenty or more and with better than sixteen years of work under all types of riders have completely unspoiled mouths. Nor do I suggest that a horse always be ridden in a snaffle rather than in a full bridle, a pelham, or a kimberwick. I have only owned one horse which I preferred to hunt in a snaffle. In dressage work four reins are needed. Horses that do not work well in a snaffle and find a pelham too strong will sometimes go best either in a hackamore or in a kimberwick. But until the young horse has learned to extend his gaits with his head low or in a natural position and to "lean" slightly on the bit as he increases his speed relying on the support of the rider's hands, he should be ridden only in a snaffle. The theory that a rider can do no harm to a horse if he uses only a snaffle is not a sound one. To get away from the discomfort of being jerked by the beginner learning to jump the horse will tend to throw his head up to avoid the pain on the corners of his mouth. If a running martingale is used—the ordinary cure—then the snaffle works on the bars of the mouth and jerking will cause pain and heavy hands callous the tender bars.

One last thing about bits. I receive many letters from young riders who have horses that will not open their mouths to be bridled, but clamp their teeth and throw up their heads as soon as they see the bridle produced.

To begin with I suggest rubbing molasses on the bit and letting the horse lick it. This will take away his distaste for it. Putting the bridle on requires a special technique. Once learned this is easy but, like everything requiring coordination and manual dexterity, it takes a bit of practice.

First slip the reins over the horse's neck allowing them to rest just behind the ears and over the poll. This will give you some measure of control for should the horse try to get away you can then grasp the reins just under his throat.

Next take the crownpiece of the bridle in your right hand and slip the whole thing into place with the bit coming up against the front teeth. Your hand holding the crownpiece will be just in front of the poll and the tension must be kept with this hand so that the bridle does not flop around.

Now cup the horse's muzzle in your left hand, the bit resting on the palm of your hand, your thumb on the lips outside the bars on the left side of the horse's mouth, and your fingers on the lips on the opposite side. You now slide your fingers into his mouth and let them rest directly on the horse's bars on the far side. Since there are no teeth there he cannot possibly bite you. And since he detests the taste of human flesh he will open his mouth at once. With your *right hand* which holds the crownpiece you pull the bridle up and the bit will slip into his mouth. Let me emphasize again that it is your right hand which pulls the bit up and into place, your left hand keeps the bit in front of his teeth, and putting and keeping your fingers (some people prefer to use their thumb on the near side instead) on the bars keeps the horse's mouth open.

When dealing with a headshy horse or with one that is too tall for the owner, grasp the two cheek pieces instead of the crownpiece in your right hand about six inches above the bit. Now, instead of taking your hand up to the ears, put it, with the cheek straps in it, on the front of the nasal bone high enough so that the bit is against the teeth waiting to be pulled into position. In every case, but especially with the headshy horse or one that objects to the bit, you should stand opposite the horse's shoulder, never in front of him.

The Teeth. The teeth are all-important to the health of the horse for unless he chews his food well he will not thrive. This he cannot do unless his teeth are kept in good condition. Let us first see exactly how the horse uses his teeth.

The incisors, or front teeth, are used to tear off the grass or to pull foliage off the trees. He has six in the upper jaw and six in the lower. These teeth should meet evenly. Some horses are born with "parrot mouths," the upper teeth overlapping the lower ones. Although the horse may get along pretty well, the teeth will not wear evenly and will have to be filed down or "floated" at frequent intervals. In male horses there are

A horse with "parrot mouth."

four teeth called "tushes" which come in just behind the corner incisors, one in each jaw.

The molars or grinding teeth lie along the rear of the upper and lower jaws, twelve in each. Thus the male horse has 40 teeth, the female 36.

The upper jaws are spread slightly further apart than the lower jaws. This permits the horse to chew with a grinding motion so that the food can be well masticated. And since the teeth are not exactly opposite each other, the lower ones being slightly inside the upper, and since this grinding wears the teeth away, it follows that they do not wear evenly; the outside edges of the upper molars are not worn away and so develop sharp points which irritate the inside of the cheeks of the horse. The inside edges of the lower molars may also develop points which can irritate the tongue. The result is that the horse, finding chewing painful, does not grind up his food as he should; hence, he does not digest it perfectly and so does not thrive. All horsemen understand this and check their horses' teeth at least once or twice a year. This is easily done by running the hand inside the cheeks, the palm against the cheeks; then the points, if there are any, will be felt against the back of the hand and fingers. The condition is easily corrected by having the veterinarian rasp the teeth. The horse does not feel any pain in his teeth and though he may be worried at the idea he should not give too much trouble. If he is extremely headshy or nervous the veterinarian will suggest a tranquilizer.

Occasionally a horse will grow extra little rudimentary teeth not much bigger than the milk teeth of a child. These are known as "wolf teeth" and appear just in front of the molars on the bars. They interfere with the bit and annoy the horse and, since the roots are close to the nerves of the eye, some trainers believe that if not removed they may interfere with his sight.

The horse's teeth continue to grow as they are worn away much as do your fingernails. Thus, should your horse have the misfortune to lose a tooth the corresponding one above or below it, as the case may be, should be removed or it will grow so long that the horse will not be able to close his mouth. This is extremely important for if this happens the horse will be unable to chew and will die of malnutrition. Should a young horse break a permanent tooth do not worry for eventually it will grow out again.

How to Tell a Horse's Age by his Teeth. It is easy to tell the exact age of the horse up to his eighth year by examining his lower incisors. Within ten days of birth the first milk teeth appear. These are the two center incisors on each jaw. They are much smaller than the later, permanent teeth though on the tiny jaw of the foal they appear to be extremely large. At four to six weeks four more teeth appear, one on each side of the center incisors. These are known as the lateral incisors. The third pair, one on each side of the laterals and known as the corner incisors, appears between six and nine months.

At two and a half years the central milk incisors are replaced by the permanent incisors; these are much bigger and easily identified. The lateral permanent incisors come into wear at four years and the permanent corner teeth at five. The horse is now said to have a "full mouth." Looking down at the incisors it can be seen that the enamel edges are higher than the centers. The center depressions are cup-shaped and, in fact, are known as "the cups." They are much darker, almost black in color as compared to the ivory color of the enamel edges. Wear gradually levels off the outside enamel. At six years old the cups of the central incisors have disappeared; the whole wearing surface of these teeth is

Telling a horse's age by his teeth.

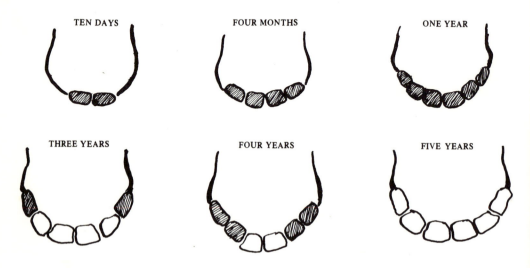

TEN DAYS   FOUR MONTHS   ONE YEAR

THREE YEARS   FOUR YEARS   FIVE YEARS

now smooth, the cups being only slightly different in color from the edges.

At seven the lateral incisors lose their cups and at eight the corner incisors are smooth. The horse is now said to have a "smooth mouth" and to be "aged." The upper incisors wear considerably more slowly than do the lower and are not used in judging age.

From nine on the exact age of the horse is more difficult to determine. The best clue is the appearance of a distinct groove which develops on the outside enamel of the corner teeth. It is called *Galvayne's groove* and appears at the age of nine at the gum line at the center of the upper-corner incisors. As the horse ages it extends down the tooth. At fifteen years it is halfway down and at twenty it extends the full length. Starting at the top it then gradually disappears until, should he he live to be thirty, the horse will no longer have a Galvayne's groove.

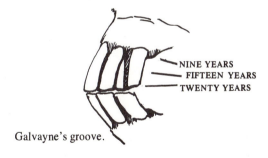

NINE YEARS
FIFTEEN YEARS
TWENTY YEARS

Galvayne's groove.

Other indications of increasing age are the lengthening of the teeth partly due to the shrinking away of the gums and the change in their angle. These things are best interpreted by the veterinarian.

Incidentally, those who have had much to do with less-than-reputable horse-dealers will have noticed that horses whose cups have disappeared are always described as being "about eight years old" since the seller knows well that it is difficult after that to tell the exact age.

Many a horse has had to be put down because his teeth have finally ground away. Shoebutton, our well-loved little Shetland stallion that we knew first as a two-year-old from Dr. Elliot and who served not only as companion but as nurse to our children, lived to the grand old age of thirty-two when he had to be put down because of his teeth. The molars were worn completely away to the gum line so that he could no longer grind his food at all.

I have since heard of a dentist who, on being faced with the same problem, capped his pony's teeth, not once but twice, and when last heard from the little fellow was still alive at 45.

Fitting a Noseband. The nose, nasal passages, and breathing apparatus have already been covered in the chapters on the origin of the horse and on the systems of the body. The *cheek bones* are worth mentioning as it is by using them as a guide that the noseband of the cavesson is fitted. The general rule is the width of two or three fingers (depending on the size of the animal's head) between the bottom of the cheekbone and the top edge of the noseband. This does not apply to the "dropped" noseband which is fitted very low to give control.

The Eyes and How They Function. The knowledge of how a horse's eyes function is extremely important to the horseman for this very definitely affects the way he handles, trains, and works around his horse.

To begin with, the horse's eyes are set very far apart and are on the sides of the skull instead of in the front of it. The horse focuses not as man does by altering the shape of the lens and so permitting the light to strike the retina correctly whether the object on which he is focusing is nearby or far away; rather he focuses by changing the position of his head, raising, lowering, or tilting it. It can thus be seen that to change focus and so see clearly a horse must have freedom of his head.

Let us first examine how the position of his eyes affects his vision. Man has what is called bifocal vision. This is possible because his eyes are set quite close together and because he focuses by altering the shape of the lens. The impressions which he sees with each separate eye are superimposed one on the other on the retinas and this single image is transferred to the brain via the optic nerves. The result is a single picture which is three-dimensional making it easy to perceive depth and the distance between objects. The horse does not have bifocal vision. Each eye produces its own image on the retina and these not being superimposed, are directed to the brain as two separate images which lack the depth of the bifocal image giving a flat, panoramic effect. However, because of the position of the eyes the horse has tremendous peripheral vision.

It will be noted that carnivorous animals which must stalk their prey have bifocal vision, i.e., eyes set close together, whereas herbivorous animals, most insects, birds and fish, and some reptiles whose main preoccupation is that of spotting distant enemies, depend on lateral and peripheral vision and so their eyes are set on the sides of their heads. A chicken, for example, when trying to focus on a worm hole will cock its head until the eye is almost parallel to the ground in its effort to see clearly.

To experience the type of double vision which the horse has, look fixedly at some small object on a wall such as a light fixture or a picture. Now press the lower corner of one or both eyes with your finger, pushing against the lower lid. This will eliminate the bifocal vision and you will see two images.

To test the absence of depth perception caused by this lack of bifocal vision, close one eye and reach quickly for an object on a table with the idea of picking it up. Nine times out of ten you will miss by several inches.

Let us see how the horse's lateral and peripheral vision affect the way we must work around him. The fact that his eyes bulge and are not set into his head enable him to distinguish movement of objects that are not only in front of and behind him but above and below him. He may not see them very clearly but unless his head is held in a line with his body (which seldom occurs) the horse will know at once if something moves directly behind him. A good horseman, knowing this, never makes a sudden movement such as waving a stick or tossing an object when he is anywhere near a horse. Young riders should have this point explained to them very thoroughly the first time they come for their instruction, for they may easily frighten a horse or a whole row of horses standing in an area with a group of beginners aboard by flapping a stable blanket or waving a switch while standing directly behind them.

Let us now see exactly how the horse focuses. When the horse's head is held high, the light comes in from below and, following the focal plane which is shaped like a ramp, is carried to the upper part of the retina. He can now see distant objects clearly but cannot see what is on the ground at his feet. When he lowers his head the light coming from above travels down the ramp of the focal plane until it strikes the lower part of the retina and the horse can see things on the ground. Thus, starting toward a jump, say four feet high, the horse will carry his head high enough so that he will see it clearly, but as he approaches the whole jump will appear hazy and he will not see the takeoff at all. It is for this reason that an unschooled horse will almost invariably come in too close to the jump, and since he cannot adjust to the rapidly changing position of the object in relation to the way the light enters his eyes, he will usually jump much too high in order to be on the safe side. If the jump is low and he is allowed complete freedom of movement of his head and neck, he will carry his head very low, in which case things above him will be cut off from his vision. The horse, with practice, adapts to this and learns to judge both the height of the obstacle and the point of takeoff even though he cannot see either one very clearly, if at all. This is not as impossible as it seems; approaching a curb you automatically make a mental note of its position and, as a rule, don't look down at it as you step up. But no one is born with this aptitude.

The lesson to be learned from this is that in schooling young horses over jumps and in introducing strange and unusual obstacles, the animal should be allowed to examine each obstacle carefully even to the point of walking him up to it and allowing him to sniff it before being asked to jump it.

Of course, the highly trained jumper will depend on his rider to make such decisions as when to take off, leads on landing, speed and angle at which he takes the obstacles, etc., but this comes only after many hours of training, training which is useless unless the horse is ridden by an equally well-educated rider.

Let us see another effect which two-image vision has on the behavior of the horse. Hold your index finger upright at arm's length and in line with your nose. Now close one eye focusing on the finger with the other. Next close that eye and open the first one. Repeat this a dozen times very quickly and your finger will appear to jump several inches from side to side. As we know, the horse's eyes are placed considerably further apart than man's. Now consider what happens when a rider comes upon a small piece of paper lying in the road ahead of him. The horse may catch sight of it with one eye only; then, as he approaches, the other eye sees it. The paper will appear to jump not just a few inches but a foot or more! This is why horses so often shy at small and seemingly innocuous objects. Knowing this the expert horseman, especially when riding a young or nervous animal, will be constantly on the alert for such objects so that he may soothe the horse and keep him moving steadily along. Failure to do so is the cause of many bad accidents. Only the other day, I heard of a man who had ridden a great deal, but was evidently still something of a neophyte and failed to notice a plastic bag which lay on the ground ahead of him. As they approached, a gust of wind made the plastic move and his horse plunged and reared. He lost his balance and tried to hold on by his reins. The horse came over with him and he suffered a badly broken ankle. Why? First, because he had not noticed the bag himself and so been prepared for the horse to shy; and secondly, because his seat was not firm enough so that in this emergency he depended on his hands and reins to stay in the saddle, thus throwing the horse off balance. Like virtually every accident the fault was in the rider's lack of knowledge, skill, and experience. We cannot blame the horse who reacted in a perfectly normal fashion.

There is still a good deal of controversy on the colorblindness of animals. For many years it was stated that except for apes and monkeys, animals were completely color-blind. Then experiments with horses and also with bees appeared to show that under certain circumstances they could distinguish some if not all colors. Whether these colors appear to the horse as they do to man is still not entirely understood. Two different people with normal eyesight will sometimes see the same color differently. But whether the horse actually sees color as color or whether different colors simply vary one from the other in value or otherwise we do not know. There have been several well-authenticated instances in which a horse has consistently shown fear of one specific color and not of others.

My own experience in this relates to Squirrel, the son of Korosko B., who was so gentle. The first time I took Squirrel out on the road as a three-year-old he had a most traumatic experience. A town road construction gang was throwing sand on some newly patched holes in the macadam. They were working from a large yellow truck which was parked on one side of the road. The men did not notice us and just as we were passing one of them threw a shovelful of sand almost into Squirrel's face scaring him out of his wits. Squirrel lived to the ripe old age of 25 and was thoroughly reliable, but he never got over his fear of yellow trucks which were standing still. He did not mind them if they were moving, he was not afraid of red, white, or blue trucks parked by the side of the road but let us come upon a stationary yellow truck and Squirrel would begin to tremble and do his best to avoid passing it.

One belief pertaining to the eyes which is completely false is the one which asserts that a horse is untrustworthy if the whites of his eyes show. This phenomenon is a purely physical one relating to whether the lids open normally or whether they open a little wider than normal, exposing some of the white of the eyeball.

The most trustworthy horse we ever owned was Meadow Lark, eldest son of Bonny.

The first time Lark demonstrated his reliability was as a yearling. Being more than half Thoroughbred he was very sensitive and sometimes playful but he had a perfect disposition both with other horses and with humans. He was also highly intelligent and later showed great learning aptitude.

The event to which I am referring occurred on a day when we were giving an exhibition for the parents and friends of the members of the riding school. This was taking place in our indoor riding hall. When it was over I came outside. Adjoining the hall was a small paddock in which Lark and a fellow yearling had been turned out for the afternoon. To my horror I saw that one of the visitors, mistaking Lark for a pony, had placed his three-year-old son on his back and was standing ten feet away taking a picture! And the yearling, though of course he had never had so much as a light pad on his back before, was standing perfectly still waiting for this strange burden to be removed so that he could go about his business of hunting out a few sprigs of spring grass!

The second occurrence happened two years later and was even more remarkable. Lark, at this time, had had his basic training. Under a good rider he might bounce a bit from sheer good spirits, especially in cold weather, but under an intermediate rider he was perfectly reliable—and very popular. This instinct to protect the young whether human or equine is very pronounced in some horses and will be discussed in more detail later.

Our students were all taught to tack up their own mounts. In adjusting the girth it had been explained to them that when the saddle is first put on, the horse will often brace his muscles in fear of its being tightened too much. Then, when he moves out and the muscles relax, the girth will have to be rechecked and perhaps pulled up another hole or so.

On the day to which I refer the little girl to whom Lark had been assigned came a bit late. She did not notice that someone had been using the colt's saddle on a larger animal necessitating the use of a longer girth and that this was the girth which was still on the saddle. She pulled it up to the top hole, the normal one when Lark had his own girth, mounted and joined the line of riders supposing that because it was at the top hole it would not need more adjustment.

In a few minutes the class took up a trot. As they rounded the corner our young rider put too much weight on her inside stirrup, and the saddle turned completely under the belly of the colt, planting the child on the ground, her head between his back feet and her feet between his front feet.

This was not the first time that I had seen this happen and let me assure you that it is one of the most terrifying things that can happen to a horse. It has the same effect on them as the tight bucking strap used in rodeos to induce the horse to buck. The normal reaction is for the horse to go hysterical with fright, taking off at a gallop and kicking and plunging in an effort to get rid of it. Yet Lark, this young, sensitive, only partly trained three-year-old with his hot-blooded Thoroughbred ancestry and his eyes completely circled with white, froze in his tracks! He remained absolutely immobile while I walked down the 150-foot length of the hall, got the child out from under him and then, undoing the girth with difficulty, removed the offending saddle. When we remember his age and the fact that he did not give in to his deeply embedded primeval instinct to fly in the face of pain or fright allowing the much rarer and weaker instinct of protecting the young to prevail, we cannot admit that because a horse shows the white around his eyes he is necessarily mean and unreliable.

Let us now discuss night vision in horses as compared to humans. Most authorities agree that, like dogs and other animals, the horse sees better at night than does a man. I have read one authority who claims that this is not so and that there is little difference in the ability of the one to see better at night than the other. Logically this does not make sense, for a defenseless animal which did not develop good night vision would have succumbed millions of years ago to the predatory instincts of the carnivorous animals which hunt mostly at night. I have also found that in riding in the dark it is far better to allow one's mount to pick his own way than to try and guide him.

However, had I been one of the doubters I had one experience in which it was the night vision of my horse which prevented a very bad and possibly fatal accident.

As every horseman knows, horses and ponies love to wander, as will be discussed in more detail in the chapter on instincts and behavior patterns. This particular incident occurred during the depression, before I had put in electric fencing. Most of our fences were low stone walls. To be sure, at night the horses were all kept inside, but we had a herd of seven or eight ponies that were in a well-fenced paddock outside the barn. In spite of the supposed security they nevertheless managed to break out now and again and one night my husband and I, returning at 2 AM from a party, entered the house just in time to hear the telephone ring. The caller turned out to be a neighbor who said that all our blankety blank animals had been roaming his fields since ten o'clock mostly at the gallop and would we come over and get them before they got into his yard and fell in his swimming pool. I had the impression that the man was more upset about the possibility of having his lawn cut up and his pool fouled should a pony go in than about any danger to the animals. And I could not entirely blame him.

It took only a few minutes to jump out of my long dress and into a pair of jodhpurs. I ran to the stable, slipped a bridle on Ariel, and hopped aboard.

The neighbor lived a quarter of a mile away and on the opposite side of the road. He had about twenty-five acres of land, most of which was kept mowed. I knew where the gaps were and was quite sure that if I could head the little devils toward home they'd not give me much trouble. But the night was as black as the inside of a huntsman's pocket. Not only was there no moon, there were no stars either for the sky was overcast.

I had some difficulty in locating the herd though I could hear them still galloping gaily. But I finally did, owing to the fact that Shoebutton and Dandy, both pintos, were among them and I could just make out the white on their bodies. They were in a field adjoining the one in which I was, so I picked up a canter and rode for a gap beyond them planning to come up on the other side and start them homeward bound. I was riding on a loose rein so that Ariel could choose her own path. Suddenly I felt the horse check for an instant, and then she gave an enormous bound and we went flying through the air. When we landed I had the feeling that I had successfully negotiated Becher's Brook on the Grand National course! I did not stop to investigate but continued after the herd of ponies which had heard me coming and were now headed directly for my neighbor's lawn and swimming pool. I finally got them headed for home and then into the paddock finding that one of them, probably Shoebutton, who was our Houdini, had managed to slip the bolt on the gate.

The next day I went back to see what it was that Ariel and I had jumped. It turned out to be an abandoned hay rake left in the middle of the gap between the two fields. If she had not seen it, or had not judged its size correctly and jumped, a very big disaster must most certainly have ensued.

Diseases of the Eye

Let us now discuss various ailments of the eye, their causes, symptoms, and treatment.

Periodic Ophthalmia. This is sometimes known as "moon blindness" because it occurs, appears to be healed, then reoccurs at more or less monthly intervals. No exact cause is known. I have heard it attributed to a blockage in the tear ducts, to a lack of vitamins in the diet of the foal or yearling, to having been raised in swamp districts, and to heredity.

This is a progressive disease which usually appears early in life. During the active periods of attack the eye waters excessively, is very sensitive to light, and becomes opaque all over. After a few days these symptoms disappear and the eye seems almost normal, but there is still a bluish tinge to it which gets worse after each attack. Sometimes there is also a spot of what appears to be yellowish pus at the back of the eye indicating an area of localized infection behind the cornea.

Eventually, perhaps after several years, the eye will be permanently covered with an opaque, milky blue curtain and the eyeball will appear shrunken. Blindness in that eye will result though he may be able to distinguish light from dark. Unfortunately, the disease usually progresses to the other eye. Previously considered both incurable and unarrestable, with the advent of antibiotics and cortisone, there is hope that if treated in its early stages it may be possible to arrest the progress of the disease so that the horse's sight will not be seriously affected. As soon as an attack occurs the horse should be put into as dark a stall as possible and the veterinarian should be called. Fortunately this disease is not contagious so there is no danger of its passing from one animal to another.

As I said, there are many opinions as to the causes for this dread disease, heredity being considered the most probable; and it is said then, that a horse that has once shown symptoms should never again be used for breeding. I can only say that our Thoroughbred stallion, Meadow Whisk, the only horse that I have personally worked with that developed periodic ophthalmia and eventually went totally blind, sired a number of colts, among them Lark and Meadow Sweet, and that none of his get ever developed it. As Lark is now twenty-six and Sweet twenty-four, I do not think it likely that they will do so.

Conjunctivitis, an inflammation of the membranes of the eye, is quite common and usually lasts only a few days if treated. It can be due to injury, a cold in the eye, or irritation due to the presence of a foreign body. There is much watering of the eye and the whites become very inflamed. A careful examination to detect the presence of a hayseed, a bit of sawdust, etc., should be followed by bathing every few hours with boric acid solution. A good ophthalmic ointment which can be recommended by your veterinarian should be applied two or three times a day to the inside of the lower lid of the affected eye. To make the boric acid solution, get boric acid crystals rather than powder. Use a pint or halfpint open-mouthed preserve jar with a clamp or screw-type lid. Boil water for ten minutes and fill the jar. Then add the crystals a tablespoon at a time stirring as you do so until the water will not absorb any more and there is a residue of crystals on the bottom of the jar when it settles. After the solution cools to body temperature or slightly warmer it is used as follows: Have a package of little balls of absorbent cotton or pull off some from a roll to be used as swabs. Have an assistant hold the horse's head steady. Dip one swab into the solution, hold the lids of the affected eye apart with the fingers of your left hand and squeeze the swab, allowing the solution to drop into the eye. Immediately discard the used cotton. Allow the horse to blink. Then repeat several times being careful never to let cotton which has come near the affected eye come into contact with the solution nor, lest both eyes should be affected, use it again on the second eye. This disease, if due to infection, is easily passed from one horse to another; if due to irritation or injury it is not so infectious. The horse should be kept in a dark stall until the eye clears up.

If a horse receives a blow on the eye from a branch snapping back in his face, he may or may not lose his sight depending on the location of the damage and on how extensive it is. Easy Going, remember, was blind in one eye when we got him. I was told that he had been burned by the spray when the stable was being disinfected with lime. I have always questioned this, as I think it more likely that both eyes would have been affected. However, he never showed any signs of being headshy, which he would have if the injury had been due to mistreatment.

Injuries to the Eye Which Cannot Be Detected on First Sight. Horses are subject to what is known as a "detached retina" and also to paralysis of the optic nerve. There is no observable symptom in either case. The eye appears perfectly clear, there is no sensitivity to light nor any discoloration, yet the animal may be completely blind. Furthermore, if he has had the ailment for some time, he may have adjusted to his affliction and behave perfectly normally. He will prick up his ears and turn toward an approaching person or animal and will avoid running into obstacles. Under saddle, he will react normally, turning his corners when he reaches them and the rider may be completely unaware the horse he is riding is blind. The handicap may go undetected until someone notices that the horse has taken to stumbling on rough ground and perhaps hitting himself against low obstacles.

What has occurred is that the horse has developed the type of extra sensory perception that blind humans develop, namely, to sense the location of solid objects as he approaches them. The seat of this perception is the forehead—the blind person will experience a definite sensation in his forehead as he approaches a wall, for example, and will receive it in plenty of time to avoid a collision.

I had a little experience with this myself some years ago. I was out on a very dark night in the Adirondacks at the camp of which I was then a director. I was on my way back to my cabin after checking on the campers. I had forgotten to bring my flashlight but thought I knew the path well enough to do without it. However, I managed to stray off the track and found myself blundering around in the woods. As I made my way slowly along in what I took to be the right direction, I found myself sensing my approach to large tree trunks, becoming aware of being close to them by a completely indescribable sensation in my forehead. It was some years later that I learned that this was a perception commonly developed by the blind.

I have had two horses which developed this ESP. One was Meadow Whisk, victim of periodic ophthalmia and completely blinded by the age of seven. He had his own private paddock which was surrounded by a post-and-rail fence. How often I have stood and watched him as he galloped around it, always turning in time at the corners. And how often I have seen him throw up his head when a mare in a nearby pasture nickered, then gallop straight toward the fence on the side nearest her, always pulling up just in time to avoid running into it. For a moment he would stand alert, then tossing his beautiful head and carrying his tail over his back he would prance back and forth along the fence with the high, cadenced steps of the *passage* while he called back to her.

The other was a pony named Popover. He got his name because he had a compulsive and most unusual rolling pattern. Take off his saddle and bridle and turn him loose either in the pasture or in the ring and his knees would buckle and down he would go. Then he would proceed to flop backward and forward from one side to the other, not just once or twice, which is normal and to be expected, but over and over again, until someone made him stand up again. If the old saying that a horse is worth a hundred dollars for every time he can roll over is true, Popover must have been worth thousands.

He was a lovely little fellow, being about twelve hands high, a skewbald with the light brown spots clearly defined and the white very white. He had a delicate head of the Welsh type and beautiful conformation, and once we had learned that he preferred being ridden in a hackamore to being ridden in a bit, especially if the rider's hands were not all that might be desired, he was a perfect mount both for beginners and for advanced riders.

He came into the stable as a six-year-old and we were delighted to discover that not only did he work extraordinarily well on the flat, but he had a great aptitude for jumping as well, and either on home grounds or in the show ring would take anything that was put in front of him. We noticed that he did seem to stumble a bit on the trail but we were doing very little trail riding at this time of year and put it down to his feet needing trimming.

In the fall of the second year that he was with us he had been jumping so well that we entered him in Madison Square Garden in a class for pony jumpers. About two weeks before the event, for the first time that I could remember, Popover brushed against the side of a wing while being schooled. He cleared the jump perfectly so I took it for granted that it was the fault of the rider who had brought him in at a bad angle. Several days later the same thing occurred. Although his eyes appeared perfectly normal I used the well-known test of waving a hand slightly behind the eye and Popover did not blink. The same thing happened when I tested the other eye. I could hardly believe it. Was it possible that this pony whose eyes seemed as clear and intelligent as any I had ever seen, who seemed to be alert and to move and jump as though he could see fully as well as his fellows, was stone blind? The veterinarian confirmed my diagnosis and said it was due either to a paralyzed optic nerve or to a detached retina, for neither of which there is a cure.

I knew that there would always be a danger of his stumbling badly and causing an accident, and that I could not therefore go on using him in the classes. I most certainly would not sell him for he might pass afterward to an owner who had not been forewarned. Had he only been a mare or stallion I would have kept him for breeding. Nor did I know anyone at that time who would be able to use him safely and would guarantee never to let him go to another home. So the poor fellow had to be put down when normally he would have had at least another fifteen years of useful life.

The Ears and Hearing. As is the case with most animals, the horse has much better ability to hear than does man. Not only is it more acute, but the range of pitch audible to various types of animals is far wider than is that of the human being. It is not known exactly how the hearing of the horse compares with that of the dog, for example, but since for millions of years hearing was one of the primeval horse's most valuable senses, it is logical to suppose that it became well developed.

Closely allied to the ability to hear with the normal hearing apparatus is the ability to "hear" through the transmission of vibrations in the earth via the bones of the feet and forelegs which transfer them to the skull and thence to the middle and inner ear, where they are telegraphed to the brain through the nerve endings of the inner ear. Since all sound is merely a matter of vibrations, it can be said that in such circumstances the horse's legs and feet simply become extensions of his inner hearing apparatus.

The ears, which are his outer hearing apparatus, are well suited to their role. Being funnel-shaped they augment the sounds they receive as does the old fashioned ear-trumpet, and being very flexible they are able to swivel in every direction thus pinpointing the direction from which the sound comes.

The ears of the more highly bred horses with their thin skin and fine hair are more sensitive than are those of the draft breed whose "ear trumpets" are quite often filled with heavy, fuzzy hair. As a result highly bred horses pick up more sounds and are more active and alert than the coarser breeds.

The ears are also used to express emotions. When they go flat back it indicates that the horse is annoyed at something. When they are relaxed the horse is also relaxed and when they prick forward he is alert and interested in something ahead of him.

In buying a horse a good horseman keeps an eye on the ears of his prospective purchase. Does he put his ears back when a hand is run over his flanks or when the buyer approaches his head? If so he may turn out to be difficult to handle in the stall or hard to shoe. When ridden in company do his ears go back when a companion is ridden beside him or approaches too closely from behind? This may indicate that he doesn't like his fellows and will give trouble when ridden in company.

Incidently, it has been my experience that the horse that has an aggressive attitude toward other horses will rarely be aggressive toward man, and vice versa. Nor have I ever found a satisfactory explanation for this.

While being examined, do his ears come forward immediately if someone makes a sudden movement or if there is a slight but unusual sound? This indicates alertness and intelligence. As he is approached from the front does he tend to shy away with his ears back or does he push out his muzzle with his ears pricked forward? The latter indicates confidence in humans; the former may indicate that he has been mistreated. Are his ears relaxed at all times while being handled, saddled, mounted, and ridden? Then he will probably have good stable manners and not be unduly sensitive under the hands of a beginner.

Besides being able to judge the emotional character, disposition, and, to some extent, background of his horse by the way his ears react, the horseman is also aware of the role of the ears in telegraphing a horse's intentions while under saddle. For example, before shying, a horse will almost invariably bring his ears forward and before kicking or biting he will lay them back. The expert horseman has plenty of time to interpret the signals and to influence the behavior of his mount. Of course, there is a wide disparity among different horses in the quickness of the reaction. We had one animal whom we christened "Long Acres" because his back seemed to go on interminably. Whether this had anything to do with the time it took a message to travel from his brain to whatever muscle was needed to carry it out I do not know. I do know that I could sense the impending action—a kick, or a shy, or an increase in gait—so long before it actually occurred that it was laughable. Needless to say he was neither very intelligent nor alert, but was perfect for our older and less agile riders.

Now let us see what kind of sounds are most apt to disturb the horse. Although it may at first seem strange, logically enough, it is the little sounds, especially a rustling in the underbrush or the sound of dead leaves blowing that startle the average, sensitive animal for it is these that awaken the fears impressed on his consciousness in primeval days when he had to be ever alert lest an enemy creep up on him.

Of course any very loud unexpected noise such as gunfire or a blast of dynamite in the distance will bring all heads up in the pasture. But whether from long conditioning or what, a loud noise which begins softly in the distance, increasing in loudness as it approaches, is seldom very disturbing. I noticed this particularly one season when the town was having a series of aerial photographs made. Again and again I would be out on the trail with a group of riders when a plane would come overhead, its engine roaring. Sometimes it would cut its motor for a minute or so and in that case it would be so close that we could often hear the men talking together. Then the motor would roar up again and off it would go. Yet never once did any of the horses or ponies show any fear.

Nor do they seem to be afraid of road machines, no matter how noisy.

Horses like to be talked to both while being worked on and while being groomed. The old-fashioned English groom always hissed softly through his teeth as he worked around his horse. The expert horseman makes a practice of talking to his horse not only while riding him but as he comes up to his stall and as he works with him. This accustoms the horse to associate the voice of his master with pleasant and normal activity and builds his confidence, and when something occurs to disturb him, the calm tones to which he has become accustomed will do much to quiet him.

Some people claim that a horse responds best to whispering and there is the oft told tale of an American groom who claimed to be able to charm and tame the most vicious horse just by whispering to him. He proved his point many times and became known as the "Whisperer." Eventually he turned up in England where a group of sporting gentlemen arranged for him to try his skill on a notoriously vicious Thoroughbred stallion that had reputedly killed one or more less-talented grooms. Bets were laid on who would win—man or horse.

The Whisperer agreed that he was to be locked in the stall with the animal where he would remain twenty minutes at which time the door was to be opened and it would at once become clear whether he was to be paid for his services or whether a funeral was in order. Apparently the odds ran heavily in favor of the funeral. But when the door was opened, to the astonishment of all, the stallion was supine on the ground with the "Whisperer" sitting on his ribs!

Most horses learn their voice commands while being lunged and if the words trot, halt, canter, and walk are each given with a specific inflection, the horse picks up their meaning very readily. In the cavalry, where voice commands had to be given in windy conditions to a platoon of sixteen men, often strung out in a column, it was customary to raise the pitch of the voice as the command of execution of a movement (ho, or now) was given if that movement called for an increase in gait and to drop it if it called for a halt or decrease in gait. That made it easier both for the men and for the horses. If the horse has learned to understand the voice commands before he has to learn to interpret the signals given by the rider's aids, teaching the latter becomes quite simple since at first the voice commands are used simultaneously with the still little-understood signals of legs, hands, and weight. The horse soon associates the one with the other and when this is thoroughly established the voice commands can be dropped.

Horses have a natural musical sense in that they learn to adjust their cadence to the beat of whatever tune is being played. In our school we had a P.A. system set up and found that music was a great help, especially with the beginners who were just learning to post or to sit to the canter. Thus, for a sitting trot we would play Leroy Anderson's Syncopated Clock and when we wanted the class to move out we would put on the Beer Barrel Polka.

Many horses learn to associate three-four time with cantering and will break into a canter as soon as they hear a waltz. Mr. Buttercup, the Arabian stallion who was on our dressage Quadrille Team, was one of these and when a slow waltz such as the Skater's Waltz or the Gold and Silver Waltz was played he would immediately go into his collected canter. Shoebutton would also, but he liked to canter to a march too, for he could fit his strides to the two-four time as well as the three-four.

Though all of our horses and ponies were fond of working to "canned" music no matter how loudly it was played, several of them were always very frightened by the music of live bands. And since we quite often took part in parades we had to be careful how we mounted our riders. The most nervous of all was Squirrel, ordinarily so reliable, whose only other phobia was his fear of stationary yellow trucks. When ridden in parades he would start to piaffe and sashay around at the first sound of a distant drum beat, and this though we used marches a great deal and Squirrel was well used to the sound of drums. Also if he should catch sight of a tuba even though it were silent he would be frightened.

I have two stories to tell of horses that, though obviously frightened by sudden noises, put such confidence in their riders that they were able to control themselves.

The first is about Sky Rocket, our phenomenal little jumper, who had been so badly mistreated that he was a nervous wreck when we bought him out of pity. He was so sensitive to sounds that once when I was sitting on him in the middle of the outdoor ring, reins long, someone fifty feet away scratched a match and before I could gather my reins, Rocket took one stride, cleared the inner four-foot fence and then the outer one. On the occasion in question, Rocket was on a team of jumpers showing at Madison Square Garden. It was a class limited to Junior Military Teams. It started with a parade around the arena. We had supposed that Rocket with his nervousness would give the most trouble so we placed him on the inside when the riders rode three abreast. In the middle we put a mare named Linnet who had never seemed to be especially nervous about anything except being clipped, and on the outside we put Bobolink, an old artillery horse. As was to be expected, Bobolink paid no attention to either the crowd or the band. Rocket appeared jumpy but was well under control, but Linnet, frightened almost out of her skin by the waving programs and the restless feet of the people sitting in the stands she had to pass, appeared to be about to take flight. She squatted down on her legs as she moved, and the young Cadet Officer, hampered as he was by his drawn saber, had all he could do to handle her.

All of the teams lined up at one end. The band got into formation behind them, and for a few moments there was silence while the teams were identified and presented to the audience. All of the horses, particularly Rocket, were restless but that was to be expected. Then, the command to salute was given and with a tremendous roll of drums (and the drums happened to be directly behind the New Canaan Mounted Troop team), the band struck up the National Anthem! The riders were now holding their sabers with the points upward and their hands at their chins. Only one horse in the eighteen or so did not bounce violently when that great burst of sound came forth, only one stood absolutely still until the very end, and that was little Sky Rocket! I can hardly believe that he really recognized the tune, though he may have, for of course we played it in reviews, but if not he certainly put an enormous amount of confidence in his rider!

The other occurrence happened in the town of Bridgeport, Connecticut, which is only twenty miles or so from New Canaan. The Troop had been asked to put a unit in a tremendous parade. We took twenty-five riders, plus myself, which gave us a leader, a color guard of four, five half squads which rode four abreast, and a file closer.

The Troop uniform is a royal blue tunic with scarlet trim, French blue breeches with stripes along the seams, overseas caps, the crowns being the royal blue and the bands the lighter French blue, high black boots, and yellow gloves.

I rode in the lead on a little Arabian gelding named Jive. He was sensitive with almost instantaneous reactions, as are all Arabians, but had great confidence in me. I carried a drawn saber, my right hand holding the hilt resting on my thigh, the point resting on my shoulder.

Behind me came the Color Guard, four riders mounted on matched chestnuts, the outside riders carrying guidons and the two inside riders carrying the Troop flag and the National Standard. The rest of the Troop followed, each four mounts being matched in color and the very last rider was a ten-year-old riding Shoebutton, our black and white Shetland stallion.

I had asked not to be put too close to any bands so we found ourselves directly behind a unit of mobile weapons and tanks while behind us trudged a company of scouts. All went well at first. I kept sufficient distance behind the vehicles, the last of which was a tractor pulling a trailer on which a swivelling cannon was mounted. Presently we came to an overhead bridge. It was a railroad bridge and just as I and the Color Guard were passing under it an express train went over our heads! It was going so fast we never heard it approach. I felt Jive quiver in fear, but I spoke to him quietly and not only did he not shy or bolt, he permitted me to turn him across the path of the Color Guard which had all bounded forward, ready to bolt into the vehicles ahead. But when they saw Jive standing quietly they lost some of their fear and their riders were able to control them.

Not so the horses and ponies which were approaching the bridge. To a man they whirled, lances flew in every direction, and spectators including some with baby carriages hastily jumped back as 20 horses and ponies fled to the rear! The only animal who was not disturbed in the least was little Shoebutton. He must have been in his twenties at this time and his calmness reassured the frightened horses just as Jive's calmness had reassured the Color Guard. While the riders recovered themselves and their lances, Shoebutton and his rider trotted calmly forward up to me, and the pony at least had the look of one who wonders what all the excitement is about.

I shall never forget the thrill which I felt when Jive, so excitable by nature, showed his confidence and trust, and so prevented what might have been a serious accident. Needless to say, when we were again invited to parade in Bridgeport, I sent my regrets at being unable to do so.

One sound of which virtually all horses are instinctively afraid is the sound of the horse clippers, especially when used on their more sensitive areas such as around and on their ears and their lower legs. Perhaps this brings up a primitive fear of the hissing of snakes, who knows? Most get over this fear once they have been clipped a few times and I have many times clipped Easy Going all over without even tying him and he has stood patiently throughout the ordeal with his head hanging, his eyes half closed and his ears at half mast. Linnet, on the other hand, was terrified; she was also unusually afraid of ropes or a hose on the ground. As for Rocket, somewhere along the line he had been clubbed over the head and though, as I noted before, he regained his confidence in people, we could never get near him with the clippers, so he had to go untrimmed. Nor could we ever get a twitch on him or doctor his teeth. It is fortunate that during the many years we had him, he never had an injury serious enough to require stitches.

The Poll. The horse's poll, the area between the ears, is important from the viewpoint of liability to injury. Although it is formed by the topmost point of the skull, the bone at that spot is quite fragile. We once lost one of our favorite young Thoroughbreds because she reared while under a doorway and fractured her skull at the poll. A less lethal injury, called "poll-evil" can be caused by a bruise which develops into a boil similar to the fistula to which the withers are susceptible. Fortunately, this is very rare, particularly since most horses nowadays are not so prone to hysterics when vanned and the like. The poll is also important because

When asked to move backwards this correctly schooled Western stock horse relaxes at the poll and jaw.

This horse, on being asked to back, is resisting by stiffening his poll and opening his mouth.

it's just behind this point that the horse is taught to flex. The untrained horse, when he feels the pressure on his bars which tells him to slow down or stop, braces his jaws and his spine at the poll and extends his nose. The trained horse relaxes his jaw, opens his mouth a little, and bends slightly just behind the poll bringing his nose to an almost vertical position. It takes many hours of training to teach the horse to relax at the poll on demand, but it is essential if the horse is to develop and keep a "soft mouth." Let me emphasize again that the bend must be just behind the ears, not in the vertebrae halfway down the horse's neck. In the latter case, quite often common in heavy-headed, heavy-necked individuals, the horse may overbend, bring his chin in to his chest, and "bore." Since the easiest way to break up resistance when a horse makes a determined effort to brace against the bit instead of relaxing his jaw is to use a "pulley" rein effect whereby the horse's muzzle is tilted to the side and slightly upward; when a "borer" takes over, the horseman has difficulty in using this rein effect and the horse can often continue for some time at his own speed before he can be stopped. The best way to combat it is to release the rein completely for an instant; the horse, having nothing to pull against, will raise his head, at which point the rider makes use of the pulley rein effect by grasping the left rein very short and the mane halfway up the neck in the same hand. He then raises his right hand pulling hard on the other rein in the direction of his own right shoulder and well away from his body. This changes the angle of the head and breaks up the resistance causing the horse to throw his head up and stop or turn sharply.

The Withers and Back

The withers (also called the "wither") is that point above the tops of the shoulder blades formed by the anterior dorsal vertebrae and is, in most cases, the highest point of the animal's torso and so is used to determine his height. The withers is also one of the most sensitive parts of the body since there are many nerves below the skin on this area. Horses like to nibble each other and to be scratched or patted on the withers. A horseman, approaching a young or timid horse will often stroke him on the shoulder and then on the withers to soothe him.

Because of its position, the wither is very easily bruised by a bite from another horse, a blow, or the rubbing of a badly fitted saddle. Many people do not realize that the saddle must conform to the individual conformation of the horse and that neither the pommel of the English saddle nor the pad, if one is used, should bear directly on the withers when the rider is in the saddle. There are horses with wide, flat withers (mutton-shouldered) and horses with high narrow withers (knife-withered). The former will need a saddle with a wide spread tree and a wide throat with panels sufficiently padded so that the pommel does not rest on the animal's withers or spine but does not sit so high off him that the rider feels like a bump on top of a log. The knife-withered horse may need what is called a "cutback" tree, where the line of the saddle at the pommel instead of being straight, a type of notch is cut in it so that it does not bear on the withers. If a saddle fits properly, the mounted rider should be able to slip two fingers in between the pommel and the withers. If a pad is used, it should be pushed up off the horse's back into the hollow of the saddle at the pommel before the girth is tightened, and the fit is tested by the rider slipping his fingers under the pad.

Fistulous Withers. Very often horses are rubbed raw on this very sensitive withers area by a badly fitting saddle. And even just a bruise where the skin is not broken can lead to a fistula. This is an abscess similar to a boil which, first localized on the wither, often extends down into the tissues and ligaments of the shoulder. It is exceedingly painful and unless treated in its early stages by a veterinarian may be incurable, although antibiotics have made tremendous inroads into this problem in recent years. The first symptom, if there is no abrasion, will be extreme sensitivity. The horse will sink down when he is saddled or rubbed or groomed in that area. If antibiotics treatment is initiated and if no further pressure is put on the area, the tenderness may subside, but more often a large swelling appears which then bursts spontaneously giving up a steady and continuous discharge of pus. Drainage is difficult since the opening is located at the top of the injury and gravity cannot help. The veterinarian will operate if necessary and will recommend treatment. In any case, as soon as the fistula opens, all of the hairy area including the shoulder directly below it down which the pus may drain should be kept heavily coated with vaseline to prevent burning and caking from the drainage.

I have heard it said by some that this ailment may be due to the abortion bacillus but have never had this authenticated. In such a case or with a fistula due to a bite from another horse, the owner cannot be blamed. But in a well-run stable, where horses are not mistreated and carefully selected tack correctly put on is used, such things as fistulas and saddle and girth sores should be practically unheard of.

Girth and Saddle Sores. We mentioned girth sores and saddle sores above so we may as well talk a bit more about them, their causes, and their cures. Girth sores come from a stiff, or dirty, or too-tight girth, the last especially if the horse has been out at grass for some months and his skin is not toughened in that area. The most satisfactory girth for the ordinary horse is the folded leather girth. It is put on with the fold to the front and is kept soft and clean by frequent applications of saddle soap and occasional applications of neat's-foot oil or one of the compounds recommended for the care of leather. This means that each time the horse is ridden his girth must be immediately cleaned of mud, sweat, and loose hairs.

Some riders buy fleece sleeves to put on over the ordinary girth. My experience has been that these can be irritating in hot weather and that unless washed and brushed almost daily they can become matted and hard. There are also string girths which are composed of a great number of cords stitched together and are supposed to provide coolness and prevent sores. I have never used these enough to be in favor of or against them, but I can say that in the forty years during which I have operated my stable we have never once had a saddle or girth sore nor a fistula.

The worst sorts of girths are the canvas girths, especially ones made of very stiff canvas. The edges of these are truly knifelike. It is almost impossible to clean them properly and I have seen many instances of sores caused by them.

Next in importance to type of girth is the way it is fitted. When the horse is standing naturally there should be four inches between the point where his leg joins his body and the edge of the girth. When the girth is first put on, it is tightened enough so that it feels tight at the bottom but your hand can slide between it and the horse's body just below where the saddle ends. It is a good practice in saddling up to put your hand in at this point and run it down to the brisket on both sides. This ensures that the hairs under the girth will be lying flat.

Some people who are ignorant will pull up on the girth until they nearly cut the poor horse in half, being under the mistaken impression that unless the girth is very tight the saddle will turn. Actually, a rider who rides correctly on a horse that is not mutton-shouldered does not even need a girth. I, myself, rode for three years using a saddle that had neither girth nor billet straps. And I remember admiring a young girl in a small show in Stamford who rode in a felt saddle (which is much harder to keep in place than a leather one) with no girth and competed successfully not only on the flat but in jumping and hunter classes!

Saddle sores from a saddle which is insufficiently padded, dirty or not smooth, or which rests on the horse's spine at the cantle can lay a horse up for some time. Horses that have not been ridden for a while are more susceptible to these. And some heavy riders who put all their weight on their buttocks can put sores on horses that never had them before. The sore, if it is not too deep, can be treated with blue gall remedy, which is gentian violet, and the horse must not be used until it has healed completely. Sometimes, if the abrasion is not bad, a very thick felt pad can be used with a hole cut directly over the sore which, if the pad is thick and stiff enough, will prevent the saddle rubbing on it. This, however, is not advised—better to turn the horse out and let him heal up entirely.

Another type of sore which develops under the saddle is called a "sitfast." This first shows up as a hard swelling, something like a "warble," more common in England than here, but much more sensitive. If there has been an old sore in the same area a dry skin will form which will later slough off. It is advisable to call the veterinarian, for an operation may be necessary.

Kidney Problems. The kidneys lie close to the surface of the back just behind the saddle area. They are subject to several ailments including bruising, so when riding bareback, double, or doing mounted gymnastics the rider should be careful not to sit or step on this area. There are also various infections which attack the kidneys, some due to moldy hay or bad feed, an infection after a bad case of shipping fever where congestion has settled in the kidneys, a chill, stones in the kidneys, and others. There are a few signs which will warn the owner that something is wrong. One is extreme tenderness over that area, the horse flinches whenever you run your hand over the kidneys no matter how lightly. Another is a tendency to stand in a "stretched" position with the hind legs behind the buttock point and the front feet pushed forward well in front of the chest. There may be attempts to urinate more frequently than usual and the urine may be discolored, being either lighter or darker than normal.

If any of the above symptoms are noticed, the veterinarian should be called at once, as there are specific treatments for each of the different conditions that can cause these symptoms. Meanwhile the horse should be brought in and well blanketed. He should be fed hot mashes with plenty of linseed in them. Stones in the kidneys cause the same symptoms as colic, the horse biting at his flanks and trying to roll, but with stones there may be sandy or gritty deposits in the urine. The patient should have plenty of fluids with a dessert spoon of bicarbonate of soda added to each pailful.

The Legs and Feet

The legs and feet are the most important and the most easily injured part of the horse. The front legs and feet are especially susceptible since they bear the most weight, especially when the horse gallops and when he lands after taking a jump. The old adage, "no foot no horse" is as true today as the day it was first uttered.

It is not always easy to decide by watching the horse move just where the trouble lies. Sometimes the rider will feel a little irregularity in the gait, especially at the trot. The horse may not really be limping but he will be "going short," i.e., taking a shorter stride with one foot than with the others. This irregularity is best determined by having the horse trot on a hard surface and listening to the cadence which should be a regular one—two, one—two.

Sometimes the horse is obviously limping badly but it may be hard to decide which foot is the sore one. Here the horse's head should be watched. If the head nods but the quarters remain steady, then the trouble is in the front leg, the horse nodding his head as he puts his weight on the sound leg. If the horse bobs his head and the opposite haunch at the same time, the injury is in the hind leg on the same side on which his head nods.

Sometimes the seat of the injury can be diagnosed by watching how the horse stands at rest. It is normal for a horse to rest one hind leg and put his weight on the other three. A very tired horse will also rest his opposite foreleg at the same time. But if a horse rests just one front foot, "pointing" it a little ahead of the other, trouble in that and probably in both front feet should be suspected. If in resting the foot the horse puts more weight on the heel in an effort to rest the toe, then *laminitis* (founder) should be suspected. If the horse puts his weight on his toe trying to save the heel, the trouble may be an ailment known as *navicular disease.*

The next test to locate the trouble is that of seeing if there is more heat in one part of the leg or foot than in the opposite foot. Since in both founder and navicular disease, as well as in sprained tendons, both front feet may be affected, the owner should put one hand on the wall of a front foot and the other on a hind foot. Laminitis is characterized in its early stages by intense heat plus such soreness that the horse, when brought out of the stable, tips along as though walking on eggs.

Strains or sprains are also accompanied by heat as well as by tenderness to the touch and soreness so, if there is no indication of trouble in the foot or pastern, the leg must be felt and examined inch by inch for puncture wounds, swelling, bruises, or heat. One of the purposes for running the hand all over the body and up and down the legs each day after grooming is so that injuries can be discovered and treated before they get too bad.

Let us now examine the more vulnerable points on the horse's extremities and discuss some of the injuries to which they are prone with the complete understanding that in any except the most superficial the veterinarian should be consulted.

The Foot

The Sole and Its Problems. As will be seen in the accompanying diagram, the foot has several important characteristics. The *sole* is

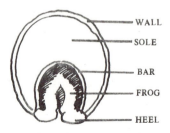

Bottom of the horse's foot.

insensitive to the touch and generally gives sufficient protection to the sensitive area which it covers. It should not be too hard and dried out for then it does not "give" enough. The commonest injury to which the sole is liable is that of being bruised. This can happen when a horse steps on the point of a sharp stone, root, or pointed piece of wood. It also happens at times that a stone will become wedged between the frog and the shoe. A good horseman makes it a practice, when riding in company, to glance every now and then at the soles of the feet of the horse traveling in front of him, as occasionally a horse will go a long way with a stone lodged in his foot and not limp enough for the rider to notice it. It is also a good idea when riding on trails to carry a folding hoof pick with which to dislodge such an object. If, however, you should find yourself away from home without this useful instrument and your horse gets a large stone so firmly wedged in his foot that you cannot free it with your fingers it is usually possible to do so by using another stone. With this second stone you hammer on the stone that is wedged on the side toward the heel of the horse; since the frog narrows and the shoe spreads toward the toe the obstacle can usually be dislodged in this fashion. If not you will have to lead your horse home as slowly as possible.

Generally, the location of the bruise can be seen as there will be a discoloration at that point. At other times, however, this is not so, and the bruised area is found by using a blacksmith's nippers and putting pressure on the sole first at one point and then at another. If the lameness is due to a stone bruise the horse will flinch violently when you hit the sore spot and try to pull away. If there is heat in the foot and the horse is very lame the shoe should be removed. The foot should then be soaked in a pail of hot water if the horse can be prevailed upon to stand in one. Following this a hot foot poultice is applied. The purpose of using continuous heat on injuries to tissues of all kinds is based on the knowledge that it is through the blood that injuries are repaired and the heat draws extra blood to the affected area.

Antiphlogistine is the poultice most often used for this purpose. It comes in a can. To prepare it the can is put in a pan of boiling water and allowed to remain there until it is thoroughly heated. The top of the can is then removed. This medication is in the form of a heavy, sticky, whitish-gray salve. Before it is applied the temperature should be tested by pulling some out of the can with the blade of a kitchen knife and holding it against the back of your hand. It should be as hot as you can possibly bear but should not burn. The sole of the foot is now completely covered with the antiphlogistine, as much being packed on as will adhere. An old gunny sack is folded into a square and placed on the ground and the horse is persuaded to put his foot down on this. So that he will not lift it up too soon an assistant should hold up the opposite front foot. A stable bandage is next wrapped around the leg beginning halfway down. It should cover the fetlock joint and the pastern and then spiral up again being tied in place with the tapes, making sure that the knots are on the outside. The bandage used should be the double-knit type but not woolen. To ensure that it goes on tightly enough, when the fetlock joint is reached the owner can give a half twist to the bandage each time it passes around the back of the foot. The purpose of this under bandage is to prevent the gunny sack, which is now going to be tied tightly just above the wall, from irritating the pastern. To tie the sack just bring up the four corners and tie them together. If the horse seems restless, a second bandage can be put on. This should come down well over the coronet so that the upper part of the sack and the knots will be under it.

Antiphlogistine, well bandaged, will usually keep hot for twenty-four hours. It should then be removed and the horse's foot checked again. If it still seems tender then the treatment should be repeated.

A pricked sole and corns are other injuries to which the sole of the foot is liable. The first comes from a careless blacksmith who has gone into the sensitive foot behind the sole, or it may be caused by a horse stepping on a nail. This causes a puncture-type wound which may not drain properly and so the horn around it will have to be pared away after which iodine should be poured in the hole. There is also the possibility of tetanus, so if the vicinity is one in which tetanus is known to exist or if horses have been imported from such a region, a preventive tetanus shot is advisable. The treatment is the same as advised for a bruised sole. When the horse has recovered and the shoe is replaced it would be wise to have the horse shod with a rubber or leather pad with tow soaked in Lysol under it. This is to prevent dirt getting into the hole and infection recurring.

A *corn* is caused by an incorrectly fitted shoe, which instead of resting only on the wall presses against the sole at the point indicated on the diagram. The shoe should be removed, the corn pared out, and three-quarter shoes which do not extend all the way back to the heel should be put on. If there is infection, poultices should be applied and the horse rested until all signs of soreness or infection are gone.

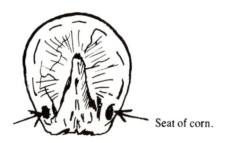

Seat of corn.

The Frog and Thrush. The frog is also insensitive but it is quite different in composition from the sole being rubbery rather than hard. The purpose of the frog is to act as a shock absorber so it is all important that the frog be encouraged to grow as thick and bulbous as possible. The horse should be shod in such a fashion that when the horse is working on soft ground the frog can be in contact with the earth; thus, instead of the full force of the concussion which occurs each time the horse's foot touches the ground in moving being taken by the heels and the edge of the wall which is under the shoe, the shock will be partly absorbed by the frog.

When the frogs are trimmed away and the horse is so shod that they never come in contact with the ground they shrivel up and become very hard. This often leads to contracted heels and navicular disease. If it is noticed that the frogs no longer appear healthy the horse's shoes should be removed and he should be put out to pasture, preferably one which is not too hard or dry, for a few weeks. Before doing this the sole should be rasped down at the heels to encourage as much pressure on the frogs as possible. If the horse has shelly, brittle walls he can have tips put on. These are light shoes which extend around the toes and stop opposite the middle of the frogs.

Thrush is the most common ailment associated with the frog. It is an infection which occurs in the center crack of the frog and is characterized by a foul odor and blackish discharge and, if unchecked, eventually lameness. Generally only the frog is affected but the disease or a similar one called Canker can spread under the sole to the point where the latter will bleed when scratched. Both these seem to be caused by a combination of bacteria and fungi which are commonly present. Some horses seem to be much more susceptible to thrush than others. I have heard thrush attributed to dirty stalls, but have known many horses that came to me from filthy stables that never developed thrush and I have also had horses kept in immaculate stalls that did develop it. I have had one experience which leads me to believe that lack of exposure to oxygen may easily be a factor, for the worst case I ever saw was the one that Bonny suffered when she was injured in the leg and so was on three legs without ever putting the other down for a long period. This case is described in detail in an earlier chapter. Since she was only hobbling and since it was very painful to her to put her bad foot down we did not try for the first month to clean out her feet. She was turned out in a nice dry pasture and it never occurred to us that thrush might develop. To our consternation, however, we discovered one day that Bonny had thrush so badly in her three good feet that the soles bled readily at any point! It seemed obvious to me that lack of oxygen was the answer since the foot which was held up was not infected.

Thrush is easy to cure as a rule, which is fortunate since it is so common. The usual remedy is copper sulfate. This is a greenish blue powder from which both green vitriol and sulfuric acid are made and thus it must be handled with great care.

The horse's foot is first cleaned out thoroughly, going well into the crack where the infection has set in. Next clean the crack again with cotton swabs making sure that there is no moisture left in it. Since copper sulfate burns very badly when mixed with water it is all important that the handler's hands be perfectly dry. The copper sulfate, which should be finely ground, is now packed into the crevice and pushed in well. A little cotton in thin wisps is packed in on top of the powder to keep it from dropping out. The best instrument to use is the back of a kitchen knife. The dressing should be removed in twenty-four hours and the treatment repeated until all symptoms have disappeared. Horses who have a tendency to thrush should have their feet picked out at least twice a day and should not be bedded on sawdust, which packs into the frog.

In cases which are very severe or which do not respond to copper sulfate, formaldehyde is often effective. This should be applied only as directed by a veterinarian as it hardens the tissues and should it be necessary to cut away the frog to treat a deeper infection the latter may have become so hard that it is almost impossible to pare it away.

Founder or Laminitis. This is an ailment which should never occur since its only cause is mistreatment or carelessness. Nevertheless it is all too common and unless it is a very mild case can never be cured entirely. A moderately bad case can sometimes be relieved and the owner may think that the damage has been completely repaired but if the horse is worked on hard roads or is jumped he will go lame again.

Founder is injury to the sensitive tissues (the laminae) which lie directly behind the wall of the hoof. These become inflamed and, since inflamed parts swell and these cannot expand due to the inflexibility of the wall, the pain is acute. A number of causes are believed to be contributory. The most common are too hard and fast work when the

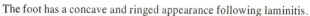

The foot has a concave and ringed appearance following laminitis.

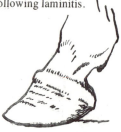

horse is out of condition, galloping and jumping on hard surfaces, drinking too much cold water while hot and then being allowed to stand, eating too much grain, foaling, and others.

The symptoms are intense pain, fever, and heat in both front feet. Later, if the case is severe, the sole drops so that it is no longer concave and characteristic ridges form on the wall as shown in the accompanying diagram. Keep the horse as still as possible and send at once for the veterinarian. He will advise treatment which may include injections of adrenalin and antihistamine, poulticing, the administration of a physic ball, hosing the feet with cold water, and a laxative diet. Some advise removing the shoes and some do not, there being two schools of thought: removing the shoes gives more freedom to the wall to spread, but it also allows the sole to drop more easily.

Prevention is the great thing and except in cases brought on by reason of extreme weight on a mare in foal, there is really no excuse for a horse being foundered if proper stable management routines are followed.

Brittle Feet. Some horses are born with such thin, brittle horn that it is hard to keep shoes on them and if they lose a shoe on the road even when walked home the foot will be badly broken. Very often the horn may grow so slowly that when the shoes are reset or when new shoes are put on the blacksmith has trouble finding an area for the nails. Some veterinarians say that lack of pigment in the horn of the foot, making it light colored, indicates horn which is often weak. It is this, no doubt, which gave rise to the old saying:

> *One white foot, buy him.*
> *Two white feet, try him.*
> *Three white feet and a white nose,*
> *Take off his hide and throw him to the crows.*

There is no doubt some truth in this but of my own experience the only horse I ever owned who had truly brittle feet was little Sky Rocket, a blood bay without a patch of white on him, whereas I have owned many chestnuts with one or more white feet, including Lark whose markings are those described in the verse, and I cannot remember any that had foot problems; and Lark is still working at 26.

Besides heredity, brittle feet are said to be due to undue exposure to water, especially salt water and possibly to bad diet when young. I would guess that the latter was Rocket's problem for when we bought him as a four-year-old he was on the point of death from mistreatment and starvation. The treatment for brittle feet includes keeping water away from the feet, painting with one of the preparations recommended for this purpose, and blistering the coronet to encourage new growth of the horn. If none of the above preparations is available, the walls can be painted twice a day with castor oil.

Sand Cracks. These are cracks in the wall which start at the coronet and lengthen downward. If not treated they spread, sand can get in, and will sometimes work its way up behind the wall to come out at the coronet. The foot may become infected and the horse may go very lame. As soon as they are noticed the blacksmith should be called and he will put clamps across the crack to hold it together and will braze the wall with a hot iron. Sand cracks usually come from an injury to the coronet. They can also be caused by rasping the wall too thin so that the secretion which keeps the hoof moist is lost; the hoof then becomes brittle and cracks. Blistering the coronet will encourage new growth. If the crack suppurates the horn will have to be cut away to allow it to drain and the foot should be treated by bathing with mild epsom salts or some other antiseptic solution.

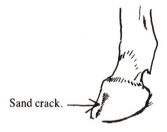

Sand crack.

Contracted Heels. This condition, which is sometimes hereditary or due to bad conformation, is very serious as it is one of the contributing causes of navicular disease. Contracted heels are also often the result of bad shoeing in which too much frog is cut away, the bars are trimmed down, and the horse is shod in such a manner that the frog never touches the ground. Contracted heels are easily identified by picking up the foot and looking at it from above. It will be noticed that the heels instead of being spread wide apart appear to be growing together and the back wall running from the heel to the coronet appears slightly concave in profile instead of straight. The frogs are nearly always dried up to practically nothing.

The best treatment is to rasp down the sole near the heels so that the frogs can touch the ground, shoe the horse with tips, and turn him out to pasture on soft, spongy ground, if possible. Sometimes it may be necessary to groove the wall of the hoof.

Injuries to the Coronet. The most common injuries to the coronet are cuts due to a horse "over-reaching" (cutting his front foot by hitting it with the toe of his back foot) or striking the inside of one foot with the shoe of the other. Normally such injuries can be treated as you would any other abrasion, namely, cleaning it throughly and using a good salve such as sulfathiazole or boric acid ointment to promote healing. However, in some cases it will develop into a *quittor* which is a fistulalike sore. Standing the foot in a pail of hot water to which a handful of epsom salts has been added followed by applying a poultice of antiphlogistine is advisable. It would be well to send for the veterinarian unless the sore starts to heal at once.

Low ring bones and side bones are bony growths on the cartilage of the coronet. They may be caused by bad natural conformation, by blows or

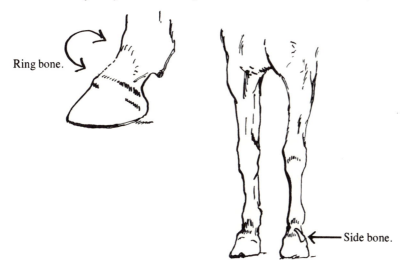

Ring bone.

Side bone.

knocks, or they may be hereditary. With low ring bones, which are found at the front of the coronet, the horse will be lame before any fever or swelling is noticed. In side bones the horse is not always lame and the lump and heat is detectable from the beginning. The treatment will depend on how severe the case is. Frequent hosing with cold water or standing the horse in a running stream will sometimes do wonders in light cases. More severe cases may require blistering. Consult the veterinarian.

The Pasterns and Associated Ailments. The natural angle of the pastern largely determines its strength. Pasterns that are straight don't have enough spring in them so the bones receive too hard a shock each time the horse moves. Those that are too long and springy can also be weak. High ring bone, which occurs at the front of the pastern midway between coronet and fetlock joint, is similar to low ringbone just described and horses with too-straight pasterns are especially prone to it.

A very common ailment, which is not serious, is called "scratches." This is similar to chapped skin in humans and occurs at the back of the pastern. Horses hunted in muddy country in cold weather are particularly susceptible. It is identified by soreness and tenderness and scabs which are formed from a discharge. The feet should be soaked in warm water with a mild disinfectant. The scabs are then loosened and picked off after which the foot is carefully dried and a thick coating of resinol ointment is applied. This ointment prevents the formation of scabs and the abrasions will heal from underneath.

Some people prefer to dust the area with an antiseptic powder (such as B.F.I. powder) but personally I have had better luck with resinol. This is also the best to use in case of *rope burns* in this area. These occur when a horse that is not used to being tethered gets the rope wrapped around his back pastern and, in struggling, burns all the skin off and often cuts into the flesh very deeply. It may also occur when a horse is tied with a rope that is too long. This type of injury is very hard to heal owing to the constant movement of the foot and the consequent continual irritation and the breaking of any scabs that form. If, however, the wound is cleaned thoroughly, dried, covered with a thick coating of resinol, which inhibits the formation of scabs, and bandaged with gauze to keep it clean, it will heal comparatively quickly. If the wound is very deep the treatment should be repeated as often as necessary until the skin has completely healed over.

Navicular Disease. This is one of the most dreaded of all ailments since nothing can be done about it, and the horse, as far as his active life goes, is doomed. It is also very common. It involves a corrosive ulcer which attacks the navicular bone (see diagram), and generally occurs in one or both front feet. It can be caused by giving the horse too much fast or strenuous work, especially asking him to take too many jumps in succession if he has been resting or is not conditioned for it, contracted heels, short, upright pasterns, shoes which have heels so high that the frog does not touch the ground, or heredity. Early symptoms are the horse pointing one front foot while resting and not resting the diagonal back foot at the same time. This is followed by mild lameness which may wear off as the horse travels, and a tendency to stumble. The horse's toes may turn in a little. The lameness will inevitably become worse, there is no cure, and the kindest thing to do is to have the horse put down.

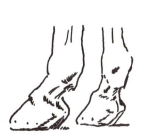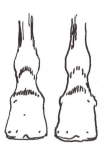

The shape of the feet due to navicular disease.

Injuries to the Bones of the Legs. As has been explained, the horse originally had four toes in front and three behind. Gradually the lateral toes became shortened until only vestiges of them remain. By examining the chart showing the skeleton of the horse and the diagram of the foot and pastern here it will be seen that the coffin bone, short pastern bone, and long pastern bone correspond to the phalanges of the fingers. The ankle joint then corresponds to the human wrist, and the hock to the

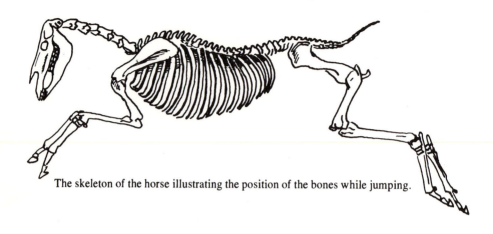

The skeleton of the horse illustrating the position of the bones while jumping.

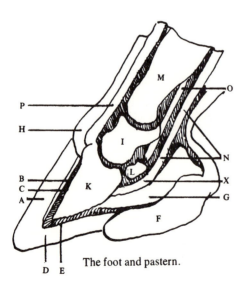

The foot and pastern.

A. CRUST OR WALL

B. INSENSITIVE LAMINAE

C. SENSITIVE LAMINAE

D. INSENSITIVE SOLE

E. SENSITIVE SOLE

F. INSENSITIVE FROG

G. SENSITIVE FROG

H. CORONARY BAND

I. SHORT PASTERN BONE

K. COFFIN BONE

L. NAVICULAR BONE

M. LONG PASTERN BONE

N. FLEXOR PERFORANS TENDON

P. EXTENSOR PEDIS TENDON

O. LONG INFERIOR SESAMOID LIGAMENT

X. SEAT OF SORENESS WHERE TENDON
PASSES OVER NAVICULAR BONE

human ankle. The cannon bones are the counterparts of the main bones leading from the wrist to the knuckle of the middle finger, in the front legs, and from the ankle to the knuckle of the middle toe, in the back legs. On each side of the cannons are the splint bones, relics of the lateral fingers. These are very delicate and subject to an injury known as a *splint*.

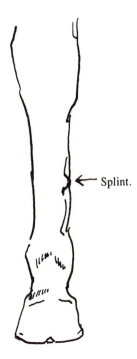

← Splint.

This is a swelling which may occur anywhere between the horse's knee and the fetlock joint. It is usually on the inside though it may appear on the outside. When it first develops, which is usually before the horse is six years old, it is soft and the horse is usually very lame. It gradually hardens, however, and the horse gets over his lameness. As the horse grows older it will often disappear completely, though not always.

If located very high under the knee it gives more trouble than if located halfway down the leg. If located too near the fetlock joint the horse may hit it with his opposite foot in which case he will go lame again. Sometimes the lameness appears before the swelling. If a splint is suspected it can sometimes be located by pinching the splint bone to see if there is tenderness in any one spot. To check, the other splint bone should also be tested as some sensitive horses will pull away simply from nervousness.

Hosing and cold-water applications can be tried. If these do not work, a mild blister such as Churchill's iodine painted on for three days, left off for three days, and repeated until the skin scruffs is sometimes effective. If this does not do the job a veterinarian should be asked to prescribe stronger measures which could include a severe blister or pinfiring.

Splints are very common in horses that are galloped carrying weight, trotted on hard roads, or jumped before their bones have hardened. The horse should be given light work while he is recovering; if he can walk comfortably he should be walked, but he should be trotted only if he does not limp.

There are seven small bones called the *carpal* bones or *carpi* which form the knee and correspond to our wrist bones. The most common injuries to horses' knees are abrasions and bruises from falls or from hitting themselves in jumping. Since there is little circulation in the knee, even a minor injury is often slow in healing. The continual action of the knee also impedes healing. Treatment can range from the application of a disinfectant such as blue gall remedy in very mild cases to wet dressings of epsom salts solution if the knee is badly swollen. If the wound is deep the veterinarian will be needed since in this case the "joint oil" may be involved and permanent stiffness result.

One large bone, the radius, forms the forearm of the horse. Because of its location it is seldom injured except from a kick or bite from another horse, which is treated like any other abrasion or wound.

At the top of the forearm is the elbow. The arm, which is enclosed in the body runs from the elbow to the point of the shoulder. Since it is so enclosed, the sideways motion is restricted and the horse cannot swing his arm out as we can.

The elbow is subject to an ailment known as a *shoe boil* or *capped elbow* which is a large abscess which forms on the point of the elbow as a result of the horse bruising it with his shoe when he is lying down. This boil can get to be as big as a small grapefruit and is very painful. The veterinarian usually lances it to let out the pus, syringes it out thoroughly with a strong disinfectant such as iodine, and recommends that it be thoroughly washed out with a suitable medication each day and that the wound not be allowed to heal until the drainage has stopped. The skin and hair under the wound and all surrounding areas should be kept heavily coated with vaseline until all drainage has stopped.

A shoe boil is not a serious ailment if treated, though it may leave the elbow slightly enlarged. It can be prevented by having the horse wear a shoe boil boot. This looks like a large black doughnut and it is strapped around the fetlock of the leg that has the boil starting. If a horse's elbows continuously have to be cleaned from matted bedding or seem slightly swollen or sore he may have the habit of lying in a position to bring on a shoe boil and it might be wise to get a boot for his protection before a real boil develops. Only very few horses, however, have this idiosyncrasy. Occasionally a horse will receive an injury in shipping which might cause a capped elbow, and too hard floors with too little bedding can also be a contributing cause.

The shoulder is vulnerable to muscular sprains and strains though often lameness attributed to this area originates in the foot. A veterinarian should be called if the lameness continues and no seat of trouble can be definitely established.

The horse's hind feet and legs are subject to the same injuries as those described above, but in addition the hock is particularly subject to two types of injuries called *spavins* and to a thickening of the back of the tendon or ligament just below the point of the hock called a *curb*. These injuries sometimes cause lameness and sometimes not, and treatment will depend on how serious they are—the veterinarian should be consulted. Another ailment known as a *Thoroughpin,* which is a distention of the tendon sheath immediately above and on each side of the joint, also occurs in the hock.

The most curious injury related to the hock with which I ever had experience occurred when a horse turned out in a field jumped over an old snake-fence. I happened to be looking at him when it happened and saw that he struck the fence with his back legs. He then took two or three strides, and suddenly went dead lame unable to do more than hobble on one back leg. There was no apparent injury other than a small scratch on the front of the hock, also a little swelling opposite it. Yet the horse continued very lame. Three or four days later I noticed pus coming out of the scratch. Further probing revealed that it was not a scratch but a deep puncture and when I pushed on the point of the lump opposite, the end of a large splinter of wood appeared! When removed it proved to be nearly five inches long, flat, and a half-inch wide at its widest with a very sharp point!

Windgalls, which are little rubbery swellings that occur on both the inside and outside above the ankles, are usually not serious and do not require treatment. They come from too hard work at too young an age or when the horse is not in condition.

Skin and Coat

To paraphrase the answer of the little boy who was asked to give the main use of cowhide, we might say that the main use of horsehide is to hold the horse together. It is also nature's most efficient way of protecting the millions of nerves which lie just under it. These nerves are not distributed evenly over the body. Those parts of the horse which are most important to him, the muzzle, nostril, area around the ears, lower legs, etc., are covered with thin skin under which lie many nerves. The wither also is very sensitive. On the other hand, the large muscles on the shoulders are much less sensitive. It is here that one often gives injections to nervous animals.

The general appearance and condition of the skin can serve as an indication of the general health of the animal. The skin should move freely and not appear to be "glued" to the tissues under it. When the latter is the case the horse is said to be "hidebound," a condition which may indicate digestive troubles or parasites such as blood worms. Even with an animal that has been turned out and is in his unclipped winter coat, the coat should not look dead or dull but should have some gloss to it.

There are a number of common ailments to which the skin is liable. Some are comparatively harmless, not infectious, and easily cured. Others are more serious.

One of the more innocuous ones is a condition sometimes called "nettlerash." Its correct name is *urticaria* and it is characterized by the sudden appearance of small raised lumps. They are not very sensitive to the touch, vary from the size of a pea to the size of a quarter, and are seen on the sides and haunches. They seldom appear around the neck or head. They are not contagious and are usually due to digestive troubles. A bran mash with epsom salts is indicated.

Similar lumps can be caused by insect bites, but in this case the horse will try to scratch them either by biting at them, kicking at them (if located under the belly), or rubbing against something.

Some horses are so allergic to insects, especially stable flies which settle on the tender skin under the belly, that they will kick themselves raw. We had such a horse named Dartmouth. He was miserable throughout the summer season and his belly was one big bloody scab. Then we discovered a repellent put out for the use of the troops fighting in jungle country in World War II. It was called *Indalone* and worked like magic. Each morning I put a little on the palm of my hand, wiped Darty's tummy, and no fly would settle. At that time we had a German Shepherd dog the ends of whose ears were similarly sensitive. We had tried every sort of repellent with no success but Indalone did the trick. I have not succeeded in finding out what medical company makes a product containing this ingredient, but if any of my readers should run across it I would be obliged if they would let me know.

A more serious affliction is *eczema* which often develops in horses that have been neglected and have been foraging on some substance that does not agree with them such as elm shoots. In dry eczema the skin is dry and scabby and there is little irritation. In the moist variety there are heavy scabs formed from a liquid which is exuded. This mats the hair in the whole area, which is exceedingly sensitive. The horse should be put on a laxative diet. The affected area should first be coated with vaseline and then bathed very gently with warm water to soften the scabs. These are then picked off. This cannot be done all at once if the condition is very bad. After bathing, more vaseline, zinc ointment, or any soothing medication recommended by the veterinarian should be applied each morning.

Ringworm is a highly contagious disease of the skin. It is caused by a fungus and both humans and animals are susceptible. It is easily identified by raised circular patches usually on the neck and shoulders. At first these are covered with hair but in a short time the hair drops off leaving the spots bare. The horse should be isolated. One set of grooming tools should be kept for his use alone and should be sterilized in a strong solution of Lysol after each use. If the body hair is long the horse should be clipped and the hair burned. Those coming in contact with the horse should also disinfect their hands after treating or grooming him.

The patches should first be washed with a warm solution of washing soda using a soft brush. Apply this to the patches only, being careful not to get it on the surrounding skin. Iodex, which is iodine in an ointment form can then be rubbed on. Use plenty of ointment and rub until the color disappears. Repeat twice a day until every spot has disappeared and the hair is beginning to grow in again.

A generally itchy skin especially in the area of the mane and tail characterized by continuous attempts on the part of the horse to scratch the areas may indicate the presence of lice. These rarely attack a horse in good condition that is groomed regularly, but are very common in animals turned out to pasture where there is little fodder, epecially in cold weather when the hair is long. The presence of lice is easily determined by parting the long hairs, because the eggs of the insects will be seen clinging to these hairs. There are various products on the market to take care of lice. Often a bath with the recommended solution of Lysol is prescribed followed by a thorough sprinkling of powder such as Pulvex. It should be remembered that this only takes care of the live insects and the treatment, especially the use of the powder, should be continued for two weeks to take care of the eggs as they hatch. The horse should be isolated and the grooming tools that are used on him should be disinfected. If possible he should be clipped and the hair burned, though it is generally not neccessary to clip his mane and tail since the hair there is coarser and it is easier to get the medication well down to the skin.

Itching alone does not always indicate that there is something wrong. Scat, a small Shetland gelding, itches more than any animal I have ever known. He is now eighteen years old and looks like a four-year-old. He is fat and his coat is so glossy that even when turned out he shines like a patent leather shoe. Yet put him where he can rub against anything and he will spend most of his time scratching himself vigorously all over, first on one side and then on the other. Tether him in the middle of an open field and he stretches the chain until it is taut, holds it steady, and rubs against it! I suspected an allergy of some sort or a special sensitivity to flies but have never been able to find bites, eruptions, or lumps of any kind. Two veterinarians have examined him and can find nothing.

Handling the Horse for Treatment

Since treating a nervous animal can be difficult, especially if the owner has had little experience, here are a few suggestions which may help. First, if possible, bring the horse into the stable to an area to which he is accustomed, where the ceiling is high enough that he cannot strike it with his head should he rear. Remove all extraneous equipment such as grooming tools, stable tools, wheelbarrows, trash cans, etc. The floor should be smooth, dry, and not slippery. There should be two stout rings attached on opposite walls or posts to which halter shanks can be fastened. The horse should be led to the area, The shanks should first be clipped to the rings on each side of his halter; then the free ends should be tied to the rings on the posts in such a manner that they can be released with one pull even when stretched taut. There is a special knot used for this in which the first part of a slip knot is tied and the loose end doubled and pulled through it but not all the way so that you have only to jerk the end and the knot is released. This is most important, for a horse may struggle and throw himself and the tension be so strong that the clips cannot be released.

There should be at least two persons present and a third can be very useful.

The following equipment, set where the horse cannot knock it over, may be needed. Absorbent cotton for cleaning the wound and a pail of warm water with a little Lysol in it; a hoof pick is generally necessary if the trouble is in the foot. If the part is to be fomented with hot compresses a pail of very hot water and two heavy towels should be used. One towel is dipped in the water and while it is being held against the injured area the other is left to soak in the hot water. When the first has started to cool it is wrung out and returned to the pail and the second is used. Epsom salts are usually added to hot water when used for this purpose. Sometimes this is followed by the application of antiphlogistine. The method of applying this is covered on page 87. At other times a wet compress is applied in the following way: After the part has been bathed as above for twenty minutes it is wiped dry. Then it is wrapped with several folds of sheet cotton. Newspaper or, better yet, plastic is put over this and held in place with adhesive tape. Next bandage with a regular track bandage which fastens with tapes or one of the type which has a special fastening which sticks to itself. The bandages should be of double-knitted cotton, not wool. If special medication is to be used this should also be at hand, and a twitch should be available should the horse require it.

Those who are not familiar with the use of the twitch sometimes think of it as a cruel means of control. Provided it is made of soft strands of cotton rope, not hemp or chain, it is not cruel. It should be used on the upper lip only, unless that is the area to be treated in which case it can be used on the lower lip, but never on the ear or the tongue.

To put a twitch on a horse, first slip his bridle on and pull the reins over his head. This will give you more control than if you try to put it on when the horse is wearing only a halter. However, with a docile horse it is often possible to get the twitch on without putting on the bridle. The assistant can hold the ends of the reins, and you should grasp them about six inches from the chin in your right hand. You should be standing on the left of the horse's head with the assistant to your left. Note that in working with a horse a person should never stand directly in front of him as he may plunge forward and can easily knock down the unwary. Both assistant and owner should be constantly aware of the horse's intentions by watching his ears.

The next step is to slip the loop of the twitch over your left wrist. Then rub his nose and muzzle gently with your left hand soothing him with your voice until he is quiet. Now quickly grasp the front of the upper lip in your left hand letting the loop come forward and encircle the part grasped.

As you hold the lip firmly the assistant quickly twists the handle of the twitch until it is tight. The horse will try to pull away at first and you will control this with the reins which you are holding in your right hand. The horse will stop fighting and will quiet down as soon as he feels the pinch, as he cannot think of two things at once. It should be emphasized that the tightness will depend on how the horse reacts. If he moves while being treated it will have to be tightened a little more.

An experienced person can put a twitch on almost any horse without assistance. In these days of tranquilizers when the horse must receive extensive and possibly painful treatment, he is usually given an injection and a twitch is not needed, except possibly when the tranquilizer is administered if the horse is particularly difficult to handle.

The twitch is often necessary when clipping a nervous horse especially around his ears, head, and lower legs, though many horses become used to clippers and no longer mind the sound after the first few experiences with it.

It is rare nowadays to find a horse that will not stand quietly for shoeing, but should you run into one of these, again the twitch is the easiest and most humane way of handling the problem. Its use for this purpose is not familiar to the average Mexican ranchero or cowboy. Instead he will collect a gang of men, throw the terrified animal who has probably never been accustomed to having his feet picked up for cleaning, hog-tie him, and shoe him. Then he will wonder why from that day forth the horse is forever terrified at the sight of a blacksmith. If asked why the simple twitch was not used his answer will be that first, it is ineffective and cruel and second, that any horse that has once been subjected to the twitch will be left with a permanently deformed upper lip!

When the twitch is removed the upper lip should be rubbed to remove the numbness and bring back the circulation and the horse should be patted and rewarded with a tidbit. To get back to the horse that is being examined or treated. As mentioned earlier, if the injury is in a front leg or foot the assistant holds up the opposite front leg. If in the back leg he holds up the front leg on the opposite side. If the treatment involves bathing for a protracted time, the owner should start with just his hand well above the injury and just stroke the horse. Then he should apply the wet towel gently in that area squeezing it to let some of the water run down onto the injury. Gradually he works his way down the leg until he can apply the towels directly over the trouble spot. The assistant should let the horse's leg down now and again and give him a rest.

If the sole of the foot is involved the assistant stands at the horse's head coaxing him with bits of carrot while the owner, facing to the rear, lifts the foot and cups the wall in his hand bending the foot at the pastern. The knee is also well bent and the person should brace himself strongly so that if the horse tries to pull away he will not succeed. In holding up a back foot, face the rear and straddle the leg as the blacksmith does, stretching it backward and cupping the wall with the pastern well bent.

A Last Bit of Advice

Keep your first-aid equipment handy in the tack room. With it should be a flashlight in case a horse gets in trouble at night and the lights are not working. A pail kept just for medication can be used to hold the cotton bandages and a bottle of Lysol for disinfecting wounds. Other remedies and ointments should be in a box together on a high shelf. Don't forget the colic medicine, capsules, and balling gun. The twitch should have its special hook where it can easily be reached. Grooming tools, especially the hoof pick, should be nearby.

V

INSTINCTIVE BEHAVIOR PATTERNS

How interesting it is to realize that with all the members of the animal kingdom that serve man, the relationship between man and his horse is unique. Man's horse has not only served him over the years as a beast of burden in his work and his wars and his explorations, but shares his pleasures as a comrade in his sport and his art as well. There are many animals that qualify as beasts of burden—the camel, the elephant, the ass, the llama, the ox. They are all strong enough to carry man and to draw his heavy loads, but they are not fitted temperamentally, nor are they sufficiently alert and agile, to be companions. Then we have such animals as the deer that have the required agility and alertness, and possibly even the temperament, but not the physical build or muscular requirements to carry weight. No! The horse is virtually the only animal the world has ever known to be both weight-carrying servant and friend.

We have inquired into how environment, climate, the lack of weapons of defense, and his herd and nomadic instincts have affected the horse's physical development and how his adjustment to these factors has permitted him to survive. Let us now see how these same factors formed his instinctive behavior patterns, patterns which, though capable of modification and of purposeful control still lie buried deep in the mentality of the horse ready to take over in times of stress. Let us look into why this knowledge is all important to us as horsemen.

The Instinct of Flight

First and foremost is the horse's "excitability to flight in the face of danger." This, as we learned, comes from his lack of ability to defend himself. Angered or frightened, the bull with his horns, the rhinoceros with his great size as well as his horn, the elephant with his lethal tusks and his tremendous weight which can crush his largest foes, the tigers and lions with their claws and tearing fangs, even the poisonous reptiles when faced with danger generally first stand and glower, then attack. But the horse at the first whiff of danger is away at a gallop.

Man takes advantage of this instinct, encouraging his mount to move out at the lightest prick of the spur, to race against his fellows just as the herd galloped across desert or plains, moving in a group.

This instinct to flight, however, brings with it a terrible fear of restraint, for if the prehistoric horse was unable to run away for any reason, he was completely vulnerable. And so today, a horse which becomes entangled in a rope or wire or one that becomes "cast" in his stall (i.e., having lain down he has rolled over and gotten into such a position that he cannot regain his feet), feeling himself unable to escape by flight, will often struggle blindly until he becomes so hysterical with fear that he may die of shock or burst a blood vessel bringing on a fatal stroke.

Many people think that this terrible fear is the sign of stupidity. Yet the instinct to run in the face of danger is the primary reason that the horse has been able to survive as a species over the last 50,000,000 years.

I have had one very tragic experience with this primeval instinct, this urge that the horse has to get to his feet at all costs. It involved a big horse of ours, a halfbred Canadian animal named Sky High. He was not especially nervous or sensitive, but somewhere in his past history he had been mistreated for he was very headshy and nervous in the stall. We had had him about eight years when I went into his box stall one morning to find him down and struggling feebly to get up. He was covered with sweat and his eyes were rolling wildly. I had no way of knowing how long he had been that way.

Now everyone should understand that a horse has great difficulty in getting to his feet if he finds himself lying in such a position that his forehand is lower than his hind-quarters. Also some horses will roll in their stalls and find themselves so close to the wall that they cannot roll back. I have even seen horses get themselves stuck this way when they rolled out of doors but too near a building.

Sky High had managed to paw a depression in the middle of his stall. Possibly he had had a mild attack of colic or possibly he just wanted to lie down and roll. Instinct tells the horse that before lying down he should paw the ground to be sure there are no enemies such as snakes around. Usually they only paw a few times, then circle around, and finally drop down. Sky, however, had pawed until there was quite a large depression, six or eight inches deep and a couple of feet across. When we found him his forehand was in this depression. It was for this reason we had not heard him struggling; the hole was in the center of his stall and not near a wall.

I realized that he was very weak from struggling and that there was danger of his heart giving out. We tried to drag him into a more favorable spot but this proved impossible. We did manage to get a sling under him and were in the process of raising him when he made one last great effort, and then it was all over. On thinking back I believe that if we had made no effort to get him on his feet at all but had tried to just keep him as quiet as possible until a veterinarian could come we might have saved him. But this was before the days when tranquilizers were common and all we could think of was to try and help him get on his feet.

A more cheerful story concerns a little mare named Pocahontas. She was a great favorite of mine for she was responsive, light as a feather, and swift. Moreover, she was so thin-skinned that the slightest touch or shift of weight produced an instant reaction. However, she was not timid and she had great confidence in people.

One morning I was informed by our cook, who took a great interest in everything that went on in the stable and who, from her strategically placed kitchen window had a good view of the buildings, paddocks, and fields, that "that thar bay mare o your'n, the one that's so skittish has been layin' in the field in the same place for the last half hour, ma'am, and eff'n I was you I'd go see what's wrong."

As I approached I could see that the mare, who was on the far side of the field, was lying in a cramped position. The reason was immediately evident. Somehow a couple of yards of square mesh wire fencing had been dropped there and the poor thing had both her hind legs thoroughly tangled in it. Now, of all the horses in the stable this one, to my mind, would have been the most apt to have struggled herself into hysterics. But quite the contrary was true. She was lying very still and had been for the past half hour, according to the cook, and there were no cuts on her legs, nor even any hair rubbed off!

I went to her head and spoke to her. She seemed relaxed. Then I examined the wire more closely. One of the strands had gotten securely wedged between the shoe and the hoof of her near hind leg. I could not possibly get it out with my fingers and probably not with pliers. Obviously wire cutters were the answer.

I went back to the stable. The stableman had gone for a load of hay and the children were in school. My husband had long since taken the train to New York. There was not a soul to stand by or help. A search for the wire cutters revealed that they were not where they belonged and then I remembered that the man had said that he was planning to take them to be sharpened. The only tools available that might be of help were a cold chisel and a hammer. But would this ultrasensitive little mare allow me to hammer directly on her foot, and hammer hard enough to cut that thick wire? Further, would she allow this with no one to stand at her head and reassure her?

Indeed she did! Nor did she move at all until I had removed all the wire, gone to her head and, snapping the shank I had brought out with me onto her halter, had indicated that she could now get up.

This proved to me once and for all that no matter how sensitive he may be, an intelligent horse, who has never known anything but kind treatment and has learned to depend on people and not to fear them, will trust them so completely that this conditioning will supersede even the fundamental and enormously strong instinctive behavior pattern which dictates tremendous struggle when unable to rise and escape.

A few years later I saw the same thing happen with Bonny, but I was not so surprised for she was a very sensible animal and not so given to excitement as Pocahontas. In coming out of an open truck that was bringing her and several other horses back from a hunt she got her foot down between the tailboard and the body of the vehicle. There she hung with me standing at her head to calm her while the men went to get tools so that they could take the tailgate off. The event occurred on the road. We had had to unload in a hurry because one of the other horses had fallen in the truck. In the scramble we had forgotten to put on the two-by-four which filled the space between the tailgate (which also served as the ramp) and the body of the vehicle.

I was not surprised that Bonny was willing to trust to my judgment and stayed in a most uncomfortable position for nearly half an hour without struggling. But I wondered whether she would be willing to get back into the truck again. However, she climbed back into it with no protest. It was a different story when we got home. Bonny was not about to come down that ramp again no matter what, although I tried to show her that the two-by-four was now in place and that there was no danger. It took a lot of persuasion, but she finally solved the problem—by jumping over, two-by-four, ramp, and all!

Bonny's experiences on the truck illustrate two points. First, that the role of master-trainer must not only be one of command but also one of support in times of danger, and further, as demonstrated by the last incident, that the horse can and does associate very specific ideas, to the point where Bonny related the danger of the tailgate with getting off and not getting on. The horse's ability to associate ideas with pleasant or unpleasant events is one of the things which makes it possible for us to train him.

Closely related to this fear of restraint and the instinct of flight being the only means of salvation is the fear of stepping on anything which may damage the foot. We have examined this before, but now I should like to look at it in a little more detail and from a slightly different point of view.

There are many authenticated stories of the horse's ability to "smell" (for want of a better word) danger when asked to step on insecure footing. If the story involves a bridge which is rickety and the horse approaches near enough to test it by putting one foot on it and then draws back, there is no mystery, for he would be able to feel the slight vibration which would tell him that the structure was unsafe.

But there are other stories about people who, riding or driving at night along a familiar road and coming upon a bridge that the horse has crossed many times before, have met with the refusal of the animal even to approach the bridge in question. Generally refusal has been so vehement that the owner has wisely decided to return home a different way. On examination the next day the bridge has been found to be so unsafe, due perhaps to a recent flood, that crossing it would indeed have been disastrous.

A slightly different story about this instinct to "scent" danger in relation to crossing on familiar but temporarily unsafe footing was related to me by a Baptist pastor who has lived on Block Island for many years. It was told to him by the local resident to whom it happened.

The time was in the 1880s. On the west side of the island lies a very large harbor known as New Harbor or Great Salt Pond. In very cold winters a large part of this quite often freezes over solidly enough for the residents to make use of it as a shortcut from the north end to the center of town. This was particularly useful when heavy snow blocked the normal road.

On this occasion the pond had been firmly frozen for a week and a well-marked track, packed by the runners of the sleighs which had been using it, pointed out where the ice could be crossed in safety. There was to be a Church Social, so at seven o'clock the head of a family, a certain Mr. Rose, hitched up his old mare, Nell, and with his wife and two children set out for town. He expected no trouble. The mare was gentle and obedient and the way was a well-traveled one. Yet when they reached the track which led across Great Salt Pond Nell balked, refusing to put so much as a toe on it. The driver coaxed, then used his whip, and the mare finally gave in. They had progressed only a distance of fifty feet from shore, however, when the ice gave way. Horse and sleigh together with the occupants found themselves struggling in icy water well over their heads. The people survived but the poor old mare, being attached to the sleigh, could not free herself and was drowned.

How did this mare know that a track which had been perfectly safe the day before was no longer safe? We can only say that the instinct which the primeval horse developed, the instinct which told him not to step on land which was boggy, this instinct on which his survival depended has never been erased from his inheritance.

The lesson to be learned from this is that in trappy country where the footing is none too good it is safer to trust to your horse to pick his way than to try and guide him. And it is this same instinct which makes a horse reluctant to step on anything unusual be it only a white line on the road. Furthermore, the more sensitive and intelligent the horse, the more true this is. The sleepy, dull, beginner's mount may tread on your toe if you put it in his path while leading him; a more alert animal will not.

It is also this instinct to avoid stepping on strange objects that saves the lives of steeplechase jockeys who so often fall directly in the path of the rest of the field. Almost never will the galloping animals even in the excitement of the race land or tread on the bodies of either man or horse.

A most remarkable happening of this sort involved Pocahontas. I was riding last in line with a group of young pupils ahead of me. We were on a narrow, winding trail going at a slow canter. The youngster in front of me was not too experienced but she was on a nice reliable old pony named Flat Top who had a particularly smooth canter. He was not an especially alert animal and so he stumbled on a hidden root. As a result the young rider flopped off in the path directly in front of me. She landed on her side with her legs doubled up. Her under arm was bent at the elbow so that her hand touched her chin. Flatty had gone on. The path, as I said, was narrow with tall bayberry bushes on each side. There was no possible way for Pocahontas to clear the fallen figure or to shy to one side. When it happened she was at the point of the canter where she was about to plant one extended foreleg from which she would push off. Her other three feet were in the air. She solved the problem by placing that one foot exactly in the center of the little space between the child's bent arm and her body, and did not touch any part of the child at all.

I, myself, escaped injury in a similar situation. I was schooling a young horse bareback over a low hurdle, only eighteen inches or so in height. She was our Standardbred filly Ariel, very intelligent and sensitive but bred to race at the trot. It takes a long time to teach this breed to canter slowly for their instinct is to go into the canter from a strong trot. Nor did I, at that time, have too much experience in schooling jumpers. We came into the obstacle at a pretty good clip; the takeoff was a bit muddy, the mare slipped, went down on her knees and I was catapulted over her head. I landed on my back directly in front of the jump. I still had hold of the reins which had been pulled over the animal's head. The mare regained her feet instantly and followed over the bar. She could not possibly avoid stepping on me because of the short length of rein but she shifted her weight so cleverly that though my riding jacket was badly torn at the shoulder and over one hip she never put a bruise on me. Did she do this because she was fond of me and wanted to avoid hurting me? Indeed not! She was instinctively saving her precious feet!

How can the owner put the knowledge of this instinct to good use regarding, for example, the horse that is afraid to go up the ramp of a horse van? He can remember to keep his own feet out of the way knowing that though the horse will not step on his toe intentionally, his fear of the unsteady ramp may distract his attention, especially if he is a sluggish animal. He can remember not to turn his back on a horse and walk directly in front of him when there is danger of the horse plunging forward excitedly to escape stepping on footing of which he is afraid. In leading the horse through a rough or narrow passageway he should face the horse, take one rein in each hand six inches from the bit, and using these to control the animal move backward slowly, a step at a time watching to see that he does not bump his hips on the sides of the passageway. Above all, if he raises his own colts he should start as soon as the colt is halter broken and leads well and get him accustomed to passing through doorways, up ramps, and even up and down steps and to put his trust in these matters in his master.

The Herd Instinct

Until 5,000,000 years ago when *Pliohippus* developed a hard hoof as a weapon of defense, the herds of horses always fled from their foes. Newborn foals could gallop with the herd when only a few days old. This is why the foal has such long legs and is so active compared to the calf, for example. With the development of this new weapon a new behavior pattern evolved. The herd learned to act together when cornered or when surrounded by enemies, such as wolves that ran in packs.

In flight the mares with their foals ran in the lead, behind ran the weanlings and yearlings, and last were the young males who had not matured sufficiently to challenge the patriarch of the herd, the mature stallion that ruled it. The stallion always stationed himself as lookout when the herd was grazing, giving the alarm which set them running, and directing their path by coming up from behind or from the side. If surrounded the herd stopped and faced to the center, foals and weanlings on the inside, mares, yearlings, and colts on the outer rim facing inward prepared to use their hind hooves in defense.

It will thus be seen that the herd instinct was not so much for companionship but for self-preservation. Though the original cause is no longer important the modern horse still feels the need for his fellows. This shows up in many ways and is far stronger in some animals than in others. Meadow Sweet, for example, daughter of Bonny, has this herd instinct so strongly imprinted in her psyche that to be alone in paddock, field, or stable to her is intolerable. When she was about eight years old someone who was not aware of this peculiarity put her out in a field which was out of sight of the other pasture and went back to fetch some others. The field

Meadow Sweet, a daughter of Bonny. She is shown here at the age of eighteen on an extended trot. As of this year (1973), she is still working in the school at the age of twenty-four having refused to be retired.

was surrounded with an electric fence which the horses had all learned to respect and they would never so much as approach it let alone touch it. But to Sweet anything was better than solitude. Furthermore she figured out immediately the best way to handle the situation with least discomfort to herself. Let me explain. A horse faced with a barrier such as a stone wall or a post and rail will stand back a few feet and jump it if he thinks it is jumpable. If it is a gateway or a drawbar too high for comfort he will go to the catch of the former or to the end of the bar of the latter and, by trial and error, will try to undo the one or slide the other until it falls down. I have seen small ponies fenced in with post-and-rail fencing find a place where the ground was a little low, lie down and wriggle and squirm until they could roll out. I have also seen them take a New England stone wall like an Irish bank, landing on top, changing legs and jumping off again. And I have many times seen a horse faced with a few strands of ordinary wire try to bend it or push it until he could scramble over.

An electric fence is one single strand of wire set at the height of the point of the horse's shoulder. The horses have learned that to touch it brings a shock. Since it is hard to see most horses will not try to jump it, though there are a few exceptions. The gap through which the horse enters the enclosed field has a similar piece of wire which, hooked onto one post completes the circuit and closes the gap. It was in such a field that Sweet found herself. She didn't try any of the methods of escape described above. Instead she used the only possible technique which would have permitted her to escape without significant discomfort from the electric shock. She backed away about fifty feet, then came straight at the gap at a dead gallop hitting the wire and breaking it instantly. She was moving so fast that it is doubtful if she even felt the mild sting but if she did she showed no reaction and actually reached the stable before the person who had made the mistake of putting her out there alone was able to do so!

Sweet never outgrew this instinct. Nor did anyone ever try to put her out alone again or let her be left alone in the stable. She also developed an intense love for her own home and her own, well-known companions. At the age of 21, we decided that Sweet had earned a happy retirement. One of the young people who rode with the school had another horse which she kept at home who needed a companion. She also had a younger sister who loved to ride. So Sweet was given to them. She was loaded into their trailer and driven the twelve miles to her new home. There she had all the possible comforts—a companion, a large, comfortable box stall, paddocks and pastures with good grass, and a trained owner who cared. But Sweet was not content. She wanted to go back to her old friends. She refused to settle down, nickering and tramping back and forth in her stall or in the paddock, snatching only mouthfuls of food. At every opportunity she undid the catch of her stall or jumped the paddock fence. Did she then set out for home? Not at all. Instead she would make her way to where the trailer stood and there she would stand looking at it hopefully. Evidently Sweet had decided that as long as she had been brought in the trailer the least they could do was to take her home the same way. At the end of three weeks the owners declared themselves beaten. Sweet was losing weight, and the continual pawing and frequent escapes did not create a tranquil atmosphere. Sweet was loaded aboard the trailer and on approaching her own stable was greeted by a tremendous fanfare of neighing to which she replied loudly in kind. Now, almost 24, she is still doing her part to instruct the young riders, mostly in the quiet work of teaching them the delicate use of the aids.

We once bought a pony who had this homing instinct developed to a high degree and he escaped whenever possible, setting out for home, which was in Mt. Kisco some twenty miles away. He always ended up on the lush lawn of a large estate about five miles from our stable. This went on for a year, when we sold him to some people who lived another ten miles further away from his original home. Several months later I was much amused to meet him on the road halfway between our stable and his newest home. He was trotting along, dressed only in his halter, and still determined if he couldn't get all the way home at least to get as far as that lovely lawn!

The horse's instinct for companionship can be annoying but it can also be put to use by the knowledgeable horseman. Take, for example, the case of the instructor conducting a trail ride with a group of not very experienced riders. When he is some little distance from home one of the riders meets with a mishap, falls off, lets go his reins and his mount departs at a gallop. The instructor has only to stop the group and keep them standing quietly between the departing animal and the home stable. After a few minutes, unable to withstand the solitude the errant mount will return to his companions. The instructor by this time should have dismounted and turned his own animal over either to an assistant or to the most able of his students. Having ascertained that only hurt pride is involved as far as the rider goes (and nine times out of ten this will be the case) he now awaits the return of the runaway. When the latter appears the instructor allows him to approach and then circles around on foot and comes up on the outside, the loose horse being between him and the group. It is now a simple thing to walk up to him quietly.

In the same way if a horse takes off from the stable alone, not to seek companionship or to find his way back to a former home, but from fright, the owner has only to wait and he will come back.

How does this herd instinct affect the behavior of horses as far as their training goes? It can help or it can hinder, depending on the circumstances. A young horse learning to jump or being introduced to the strange sights to be encountered on roads and trails will be much more confident if allowed to follow a companion. On the other hand, the animal which is never trained to work by himself will inevitably become a "barn rat" refusing to go on the trail by himself unless ridden by a very competent rider, shying for the door of the arena, etc. So very early in his career, the properly educated animal learns that, instinct to the contrary, his first duty is to go where his trainer directs whether he is working alone or in company. He must learn to leave the barn or paddock without a companion just as willingly as though he were with one. He must learn to go just where his rider wants him to go rather than to pursue his own preference of returning to the stable or following a companion who may have taken a different route.

This element in equine education is often neglected by trainers who carry the training of their animals only to the point where they will carry a rider on a trail following in line behind a leader, or work in a ring the same way. How many times does a parent, having seen how well his child is doing in his riding class, buy or rent him an animal for the summer! On the very first day he learns that the pony or horse that seemed so docile when the young rider tried him out in company at the dealers refuses to leave the premises with his new owner but balks, whirls, and returns at a gallop before he has been persuaded to go more than a few hundred feet.

This habit, which the British call "napping" is especially common in ponies. Thus ponies have the reputation of being more willful and stubborn than horses. It is my contention that if horses did not receive any better training than do ponies, especially the ones that are too small for the average trainer to ride, they would be equally willful.

Let us examine this a little more closely. The only way to overcome this herd instinct is by making the animal so responsive to the commands given by the rider through the use of his aids that he obeys automatically. This means many hours of training, first on the lunge, where the young horse learns to obey the voice commands, and then under the hands of an expert trainer who teaches the lessons of immediate response to the aids by asking the horse to execute certain patterns and movements and to repeat these over and over again until they are executed automatically and smoothly. The most important of these is to move out and to change direction immediately in response to the legs. Working both alone and in company the equine pupil learns to turn to the right or left, to execute small and large circles and half circles, to change from one gait to another, to extend and shorten his strides without changing the cadence, and to do all these things whether the rest of the group is performing them at the same time or not. The first few times that he is taken on the road he usually goes with a companion but as soon as he has become accustomed to traffic and other common sights he is ridden alone. A horse trained this way will never be a "barn rat."

The wise trainer never allows an uneducated rider to handle his young horses (or his highly trained horses for that matter) except under his strict supervision. The young horse not knowing enough would not respond to uncertain aids and an uneducated rider does not know the language of the aids well enough nor does he yet have sufficient security of seat to use them properly, so such a combination—young horse, new rider—will only confuse both horse and rider. A relatively new rider should never handle a highly trained horse unsupervised because such a horse is generally so sensitive to the aids that if subjected to indifferent use of them he will become very upset and—at least temporarily—spoiled.

But what happens in the case of the pony? Generally his training consists of being mounted and ridden by a youngster who is a strong rider but not necessarily an educated rider. He is worked only in company, and when the young animal has gotten over his repugnance toward bearing weight, will move forward at the walk and trot while following in line,

Many people believe that all Shetlands are untrustworthy. Here is Dandy Beans, an over-sized (12:2 hand) Shetland from Belle Meade Farm. He is shown in a clown act in the New Canaan Mounted Troop Annual Equestrian Circus where he impersonates a taxi cab. Without anyone at his head he allows a voluminously petticoated ''lady'' to climb aboard.

and will stop, he is considered "broken" and sold at auction or to a private buyer. If he goes to a good stable or to a child with some experience, who is still studying and so can work under the direction of a trainer, well and good. If he goes to a typical "hacking" stable his education will not be enlarged and he will simply go out day after day on the trail, the riders following along one behind the other holding on to the pommels of their saddles in blissful ignorance of how little they are learning. No wonder all of these animals become "barn rats."

Yet, with proper training, ponies are no more stubborn or willful than horses and far safer for they are not inclined to be so hysterical, they are better proportioned to their riders, and a tumble off a ten- or twelve-hand pony is far less serious to the seven-year-old rider than is a fall from a fifteen- or sixteen-hand horse. Furthermore, being in proportion, the child's legs come where they should so that the animal responds as he has been taught. Also in my opinion no rider is safe nor should he be considered ready to begin his advanced training until he is as much at home without reins and stirrups as with them and until he can dismount from a horse moving at any gait landing on his feet and holding on to his reins. This is easily taught when the riders are mounted on animals of the right size.

Youngsters who are members of pony clubs receive training in schooling their mounts correctly, as do those riding with units such as the one of which I was Commandant for 25 years. Breeders of high-class ponies which will later be sold for large sums as hunters or show ponies train their animals very carefully. They also command very high prices. And of course there are a few small riding establishments where ponies are used and their training is considered part of the education of the pupils.

I was the first instructor to introduce ponies in the east. This was back in the early twenties when there was a terrific prejudice against them. In my first book, *Teaching the Young to Ride,* written in 1932 for the purpose of introducing methods of teaching children as young as four or five to ride, I said that ponies, if well trained, were invaluable. One well-known critic said that *all* ponies were necessarily willful and *none* were to be trusted. I wonder how this man feels today with the present popularity of the pony and the many beautiful animals that are as well trained and obedient as any horse that is to be seen in the hunt field, at Pony Club Rallies, and at every horse show.

At the time that I first began buying ponies I was fortunate enough to hear about the Belle Meade Farm at Front Royal, Virginia. It was run by a man named Dr. Elliot who bred and trained American Shetlands. He had a herd of several hundred and each year got out a brochure listing the animals for sale and giving their age, height, color, disposition, the extent of their training, and whether they were considered suitable for beginners. Over the years I bought at least 15 animals from him. Not one failed to live up to his description. Later I went to the farm and talked with Dr. Elliot. I was very interested in finding out about his training program. He said that since Shetlands mature earlier than other breeds, he started his yearlings in long-rein driving and then drove them to a breaking cart with himself in the driver's seat. He worked them first in an arena, next on a track which ran around it, and finally took them out on the road, driving them regularly until they responded well and had settled down. This training was followed by work under saddle using young riders whom he had trained himself and who understood the work. The ponies were worked both in company and alone and though most had only preliminary schooling and possibly some work over obstacles, they were taught to move out and to turn readily, and not one was ever a "barn rat." He was also very careful to find out the age and experience of the prospective young rider and never sold anyone an animal that would not be right for him.

To get back to the mother who buys or rents a pony for a child who is still only at the beginning of his career as a rider, it should be remembered that only one pony or horse in a hundred, no matter how well trained, will not deteriorate and develop bad habits when handled entirely by a beginner. For this reason only a reasonably knowledgeable and experienced rider—child or adult—should have a horse of his own to look after and ride without supervision.

This primary behavior pattern, the desire to stay with the herd, which is related to the homing instinct, sometimes seems illogical. Thus many people cannot understand why, for example, in cases of fire in a stable, horses that have been gotten out of the burning building will sometimes try to rush back in if not restrained. This is simply a case of hysteria and fright taking over in times of stress. The stable, being home, represents safety. In the stress of excitement the fact that it is the stable itself that is on fire is not clear to the terrified animals. They only know that there is danger and that the stable has always been synonymous with safety and with the presence of the other animals. In such an emergency, horses that are led out must be put in a paddock or field well fenced in and as far as possible from the burning building. They should never be brought out and turned loose on the assumption that they will go away; and if there is no suitable paddock some person must stand guard at the doorway.

I would like to give one more slightly different example of the herd instinct. It relates to the horse's well-known ability to find his way home, but in this case the mare in question did not try to go back to her own stable, nor, in fact, to any familiar territory; rather she took her master to where his companions had gone.

The time was during the latter part of the Civil War. My great uncle, Richard Bolling, was a Captain in the cavalry of the Confederate Army. In those days officers brought their own mounts and their own body servants when they enlisted. Captain Bolling brought with him a young mare which he had schooled himself. As were many horses used by cavalry, she had been taught to lie down on a voice or hand signal. This was so that in battle the horse could be used as a bulwark to protect the soldier, the latter lying prone behind him, his rifle resting on the body of the animal. When one realizes how terrifying to the horse must be the sounds of battle one realizes how remarkable it is that an animal so timid by nature and whose basic instinct is flight in the face of danger can be so conditioned!

Captain Bolling went through the early years of the war without harm. However, in the Battle of the Seven Pines which took place some 20 miles outside of Richmond, he was shot in the thigh and fell to the ground on the battlefield. Seven cavalry charges passed over his unconscious body, but forbidden by instinct to put a foot on a strange object, not one horse stepped on him.

The Confederates were defeated and pushed back. The Union soldiers followed them for a short distance, then bivouacked for the night. The Confederates continued on to Richmond. It must be noted that they had not come from this direction and the Squadron as such had never before been in Richmond.

My uncle regained consciousness some time after the battle was over to find his mare, who had been hiding in the nearby woods, standing beside him. This again is to be expected. He was the only familiar person or animal to which she would be drawn. He gave her the signal to lie down beside him which she obeyed.

With great difficulty, for his thigh was shattered, Captain Bolling managed to pull himself across the seat of the saddle, lying helpless on his stomach. He then lost consciousness again. When he came to himself some hours later he found himself being lifted down by men of his own Squadron. The mare, carrying his inert body, had found her way through and past the Union lines then through 20 miles of bleak and unknown territory searching out the path that eventually took her right up to the stables of her own Squadron.

Remember that the mare had never been in Richmond and that she passed by the stables of the Union horses. Nor do I believe that she consciously performed the feat with the idea of saving her master's life. Rather I think that several basic behavior patterns as well as the conditioned training which she had received affected her behavior. It was instinctive for her to retreat off the battle field and to linger in its vicinity. She was many miles and probably many months away from her normal home territory and so there was no impulsive urge to find her way back there. At the end of the battle it was instinctive for her to seek out some being whom she trusted. I do not believe that horses have the same love and loyalty toward their masters as have dogs. But, as demonstrated by Pocahontas, they can and do learn to trust and to depend on them. One should remember that she was an animal which my uncle had trained himself, not a strange one that he had acquired when he joined the army. That she should remember her training and should have lain down is also normal since this would be merely a matter of automatic response to a well-learned signal or command. It is also normal that she should try and seek out her own companions. As we know strange animals are not usually welcomed into a herd.

What is really remarkable is how she managed to pick the line of her companions and to follow it unerringly at night and through strange territory, stopping only when she reached them.

I think the only logical explanation is that she did this through the sense of smell. Possibly, also, the vibrations of earth caused by their march and carried to her through her feet, thence to her brain where they registered as sounds, may have helped her. This would be especially true if the mare found her master shortly after the battle ended so that the retreating Confederates were only a few miles ahead.

Neither this ability nor the remarkable sensitivity to odors is easily understandable to man. But that animals can follow scent in a most remarkable way is well known to anyone who has followed either foxhounds or, in particular, bloodhounds.

I have heard it said that the range of man's scenting ability as compared to that of animals may be expressed by comparing the vision of a man who can only see objects when they are no farther away than arm's length to a man with normal vision. The scenting ability of the human race is equal to the seeing ability of the first man, that of animals to the second. We also know that scent lies not only on the ground but is also held by foliage.

Nevertheless, this story is one of the most remarkable of any that I know. And that it is true you may be sure for I knew my uncle, permanently crippled by the wound, very well, and he told it to me himself.

The Instinct to Roam

Anyone who has had much to do with horses or ponies is aware of their very strong instinct to break out of the pasture and to wander in search of greener grass. This may seem to contradict the instinct to keep together and to stay in or go back to familiar surroundings. But if we examine the facts we find that they do not contradict this pattern, for a horse will never leave a companion or a herd and go off by himself—generally if one breaks out they will all go together. The only exception is with a new animal who has not identified himself with the herd, but then he is not searching for fresh pasturage but is following his homing instinct.

The instinct to roam of course goes back to the fact that in primeval days the horse was necessarily a nomad. For millions of years he roamed from territory to territory looking for forage, escaping from enemies, and so forth. Small wonder that his descendants retain this instinct. Nor is there a more troublesome one if some or all of the animals are good jumpers or if there is a Houdini in their midst who can figure out how to undo gate latches and slide bars that close gaps.

Horses are very selective in the grass they eat. They will crop certain parts of the pasture right down to the bare earth and then leave the area for another even though there is plenty of good grass left in other parts. One theory is that the grass they have refused is fouled with manure. Yet many horses will eat pure manure and this is said to indicate that their diet is lacking in certain necessary nutritive elements. Others say that they can sense the presence of poisonous weeds and will not eat the grass around them. This is undoubtedly true in many cases, but I have often examined a patch of grass which has been refused and which is right in the middle of a well-grazed paddock and have been unable to find evidence of any foreign plants whatsoever. My own observation shows that many times they prefer short grass to long grass, but, there again, there are exceptions.

Perhaps the truth of the matter is that they just get bored with their surroundings and their instinct to roam takes over. The only solution is effective fencing and experience has shown that electric fencing is both the most effective and the cheapest.

Of course if you have plenty of pastureland you can satisfy this instinct by keeping your animals in one lot until they start breaking out, then putting them into another until they graze that one down, and so on. Putting a fence down the middle of a field can also help because this ensures that after finishing the grass on one side, that area has a chance to grow back while the animals are pastured in the other.

The Protective Instinct

With the herd instinct goes the instinct to protect its members from the intrusion of strange animals, including other horses. Every experienced owner knows that the introduction of a new animal to his stable means from two days to a week of excitement. During this time the stranger goes through a period of persecution and hazing. Each time he approaches the group trying to make friends he is unmercifully chased away by the animal highest in the pecking order that happens to be present. Generally no harm is done, provided the newcomer keeps his distance. There will be flying hooves and a great deal of squealing to be sure. If there are male animals present, either geldings or stallions, and if the stranger is a male, he may be attacked in a different fashion. For whereas mares turn their tails and kick out with their hind legs, which is primarily a defensive action, male horses box with their front hooves and use their teeth, trying to grab the withers and then the forelegs in an attempt to cripple. After several days of misery the newcomer will usually find a buddy and after much blowing out and deep sucking in of breath, muzzle to muzzle, the two will pair off. Generally this new friend is the next most recently accepted member of the herd, even though the latter may have joined several weeks or months before. The result can be quite ludicrous. I remember one such strangely mated pair, the one a very large, halfbred heavyweight hunter, the other a ten-hand Shetland. The former joined the herd some six weeks after the latter and after the preliminary hazing period had passed, for as long as the two remained with us they were inseparable. It was particularly amusing to watch the "mutual admiration society" or "you scratch me, I'll scratch you" routine, which is usually done by the two horses standing facing in opposite directions and both nibbling gently at the other's withers. In this case, because of the disparity in size, the Shetland, standing in front of his friend was just able to reach partway up the latter's broad chest while the hunter performed his part of the bargain by nibbling away at the top of the little fellow's rump!

Another facet of the protective instinct is the attitude of mature animals of either sex toward foals and yearlings. Not only is the new mother herself on the defense with her little darling, it is not at all uncommon for a gelding to have a very highly developed foster parent instinct. Easy Going, mentioned earlier, was one of these. Each spring when the new crop of foals arrived he would select one to be his foster child, joining with the mother to help guard the newcomer. If Bonny happened to have a baby it was always this one that Easy chose, but if not, he'd offer his assistance to Ariel or to one of the others. The mare understood his role perfectly and though normally a new mother will be extremely protective and jealous toward her newborn, chasing every animal away at least for the first week, Easy's assistance was always accepted with pleasure. From the very first day that the new foal made his appearance out of doors, Easy would take up his position, placing himself so that the mare was always on one side of the foal and slightly ahead, and he was on the other and slightly behind. He could thus prevent the other horses coming up from behind and could keep the foal from straying or stopping to make the acquaintance of a slightly older foal, while the mare could seek out the best grazing spot and prevent excessive curiosity on the part of animals in front of her.

The stallion will never harm a foal nor a yearling and generally speaking will accept colts up to the age of two. After that they are open to attack, and if newcomers, will have to go through the same period of trial that all the other newcomers must put up with.

This protective feeling of responsibility toward the young, can in a few horses be extended to a similar protective instinct for young children. I have mentioned the case of Lark, who, as a yearling, allowed an ignorant father to put his baby son on his back and then stand off while he took his picture. This could be attributed to a generally polite and responsible disposition. But let us examine the reactions of Korosko B., our five-gaited American Saddle Horse stallion, father of Squirrel.

Bring Korosko out on a frosty morning and he would let out a trumpet call of delight, and stand still only long enough for me to get into the saddle before prancing around, nostrils flaring, executing a *piaffer* while I caught my stirrups, then arch his neck and take a few steps of the *passage*, with every appearance of being about to go into a *capriole*. However, let me reach down, grasp the hand of my two-year-old son and hoist him into the saddle in front of me and Korosko B., with no sort of intimation from me, would forget his foolishness (which he knew I enjoyed as much as he did) and walk or take up his delightfully smooth "slow gait" as gently and quietly as an old cow! I will never forget the expression on the face of a newly hired stableman the first time he saw this performance. Having stood well out of the way while Korosko was doing his best to make like the dangerous animal that most people believe all stallions to be, he thought I must have gone mad when he saw me reach down for the child. Then when he saw the change in attitude and behavior of the erstwhile ferocious animal he thought I had some magical power!

Korosko was equally careful of young children when I was not around. I came home from a trail ride some years later to be told that another son, then three years old, had fallen through a hole in the floor of the loft over the stalls through which hay was pitched.

"Whose stall?" was my first question.

"Korosko's, but he didn't do a thing, just waited until I ran down the ladder and opened the door and got Toby out," said the older son who was up in the loft with his brother at the time.

"And I didn't get deaded, either," said Toby with great satisfaction. He had been used to horses from birth and so was not at all afraid and, since he landed on the pile of hay which had just been pitched down, was not even bruised.

Korosko B. handed this trait of protectiveness toward young children along to his son Squirrel.

Squirrel was one of the dearest horses we ever owned. Not only was he careful with children but he was also careful with adults or older children who were in some way handicapped. Such was his reputation that both because of this characteristic and because he had the smoothest possible slow gait, a delightful rack if you didn't want to post, and the kind of canter which in the old days used to be described as one in which a rider could "ride all day under the shade of a tree," he was exceedingly popular with everyone. He was also the only five-gaited animal that we ever owned that was excellent in the hunting field. This came about because one of my daughters, at the age of seven, wanted a mount that she could show in Children's Hunter classes. The ponies were all too small to jump the required three-foot obstacles in the ordinary outside hunt courses. Unbeknownst to me she trained Squirrel herself, discovered that he had great aptitude as a jumper, and showed him successfully, without the judge ever realizing that he was five-gaited! Because of his beautiful disposition and his smooth gaits he was always assigned to people recovering from injuries of one kind or another, and I myself rode him until a few weeks before my youngest daughter was born and was up on him again a few weeks afterward.

Of course we did not know of Squirrel's protectiveness when we first got him, but we were soon to learn about it. He was barely two at the time and had only been with us a few weeks when he injured a hock jumping a low wall between two fields.

The injury was not noticed immediately, and when it was, infection had set in which required bathing with hot water and epsom salts followed by a poultice of antiphlogistine. I happened to be alone on the place. I cross-tied the colt, got my water, towels, and medication ready, and started to work. Since it was probably the first time he had ever had treatment of this sort and since the whole area was very tender and sore, the colt was understandably restless, swinging his quarters from side to side, pawing, and so forth. Suddenly he stood still.

"Aha!" I thought, "he's settling down because he knows I'm not going to hurt him!" Then I looked over my shoulder. Directly under the belly of the colt was standing the same little two-year-old that loved so to ride the stallion with me. He was eating an apple and had no idea that this was not the best possible place to stand especially when a young and nervous colt was being subjected to treatment which he did not understand. But Squirrel's protective instinct for the young had overcome his natural reaction of trying to escape from pain and he was standing like a rock! Needless to say as soon as I had told the boy to move away, the colt became restless again but eventually his confidence grew and I was able to treat him satisfactorily.

The average mare would never hurt another's foal, much less her own, but here again there are a few exceptions. Quip, a good-looking Thoroughbred mare who was always well-mannered toward humans was so nasty to her own foal that she bit off a piece of his ear when at the age of a month he reached into her manger for a bite of oats; but she was a rare exception. Most mares appear to be very affectionate toward their young, nuzzling them, whinnying to them, and teaching them to stand directly behind them under the shade of a tree during fly season so that the long tail of the mother can whisk away the flies that the fuzzy little appendage of the foal cannot reach.

This fondness generally disappears when the foal is weaned. In the natural state this doesn't usually occur until the new foal is expected, but in most stables, for the sake of the mother, a foal is generally weaned at the age of six months. There are always several days of loud complaining on the part of both mother and child, and they are not allowed to meet again for several weeks, by which time the mare's milk will have dried up. When they do, generally speaking, they will not be especially interested in each other, the maternal ties apparently having been broken.

But, as always, there are exceptions. One of these was Ariel. She did not have her first foal until she was about fourteen years old. From the beginning she was unusually possessive in her attitude toward him. When weaning time came she refused to settle down and each time, having kept them apart for a month, when they were allowed to meet once more she would go back to nursing him. Finally, since Ariel was expecting again, when the weanling was ten months old I sent him away to a farm to board having gotten tired of Ariel's continual efforts to break out and get back to him.

He seemed to be doing very well so I let him stay where he was until Ariel's second foal was almost six months old. The first colt, now a well-grown animal almost eighteen months old, arrived in a farm truck which drew up to unload outside the pasture gate. Ariel and her second offspring were four acres away at the other end of the pasture. As the truck stopped the colt let out a shrill whinny. Up came the mare's head and she dashed for the gate shrieking her delight all the way.

The minute he was unloaded the yearling went up to his mother showing every symptom of joy at seeing her, they sniffed noses with mutual soft murmurings of endearment and then the mare, signaling in some way unintelligible to me, indicated that a feast had been prepared for the return of the prodigal, and the gangling yearling knelt and proceeded to suckle!

The Territorial Instinct

Since the horse, by nature, is a nomad, one might conclude that the instinct to defend a specific territory would not be an inherited one. But

Richard Ardrey in his very fine book, *Territorial Imperative,* explains that in addition to the instinct to protect and keep for one's colony a definite location or territory, nomadic animals apply this instinct to moving territories. Animals prevented by man from being nomadic apply this instinct today as shown by their aggressiveness toward strange animals when they are first introduced into the herd. Ardrey also describes another type of territorial imperative inherent in most animals including man, and that is the instinct to preserve an area of open space around each individual, drawing attention to the fact that two people talking to one another will rarely face each other directly but will stand at a slight angle to each other, leaving plenty of open space between their faces. He also mentions the uneasy feeling which everyone has experienced when a person who may perhaps be very nearsighted insists on standing too close and "breathing in your face" as he vociferously expounds some favorite theory.

This same instinct to preserve intact the individual territory is well demonstrated when one watches two colts playing in a paddock or field on a frosty morning. When they first come out of the stable they will separate, each sniffing the ground until he has found a good spot. Having examined it carefully and pawed at it (the primeval behavior pattern of checking for snakes or other undesirable residents) the colts lie down and roll. Having scratched themselves luxuriously on both sides they will rise, shake, and perhaps give a buck or two. Then the play ritual will begin. It may take the form of a mock battle with both animals rearing as high as they can and boxing. Or, especially in the case of fillies, there may be much kicking and squealing. But if one watches closely it can easily be seen that the territorial imperative relating to the proprietary rights of each individual for his own personal space is well observed. The flying hooves do not quite make contact, the threatening forelegs return to the ground without having done any damage.

This instinct to play and "get the kinks out" without hurting one another can be of great use to the riding instructors whose animals must not be "too full of beans" when the beginners arrive for their lesson.

Let me first explain that horses behave very differently in winter—when the thermometer drops, the ground is frozen, and they have to be kept in the stable when not being ridden—from the way they behave in summer. In our establishment we always used to refer to our animals as having "winter" and "summer" dispositions. There were always a few even in winter which, perhaps because of age, unusually cooperative characters, or thicker hides, would submit to the uncomfortable burden of the riders whose hands were heavy and whose legs flapped and pounded their sides. But the majority of the beginner's mounts, though sensible enough in warm weather, as soon as winter came would behave like fools and pretend that there was a saber-tooth tiger hiding in every corner of the indoor riding arena.

At first we tried to take care of the situation by having experienced riders work them out for a half-hour or so previous to the lesson. But when you are faced with a beginner's class of fourteen riders coming as soon as school is out in the afternoon, one instructor with only one or two assistants has not time to work each horse individually. Then I discovered that if, instead of having the horses ridden, I simply turned them into the indoor hall in groups of five or six and encouraged them to buck, rear, play, and chase each other around for fifteen minutes or so, nearly all of them would then assume their sedate and respectable "summer" dispositions—until next time.

VI

HOW DOES THE HORSE COMMUNICATE?

How does the horse communicate?

This is a three-way question: How does the horse communicate with other horses? How does the horse communicate with man? And how does man communicate with the horse? Since we know the most about the last we shall examine it first.

Man, as we have seen, communicates his intentions to his horse through his voice, making use of both the tone and specific words. He communicates through his gestures and his attitudes. And he communicates through the "language of the aids."

The horse, being far more intuitive than man, as well as more sensitive and alert to bodily movements, learns to understand what man is trying to get across to him long before man is equally educated in interpreting what the horse is trying to say. It takes a young horse only a few lessons to learn the meaning of walk, trot, turn, halt, and canter, for example. Some learn too well. Shoebutton was one of these. In instructing a class, I might say "To take up the canter one must...." And before I got any further Shoey would have obediently picked up his little rocking-chair canter, to the surprise of the beginner on his back. I finally had to resort to

134

spelling such words when Shoey was in the ring excepting when I was ready to have the class execute the command!

They also learn phrases. Horses used in trail work quickly learn "catch up" and "pull to the right, you're out of line," for example. Sweetheart was a very trustworthy beginner's mount. I was once at the head of a line of beginners as we were returning to the stable when a car came up behind us. I looked back and saw that Sweetheart was out of line.

"Pull your right rein, Johnny, and get back into line; you're in the middle of the road and a car wants to pass," I called. Johnny, aged six, pulled manfully on the left rein instead, but Sweetheart moved to the right and got back where she knew she belonged. Johnny may not have known his right from his left but Sweetheart knew what "get into line" meant!

Any horse can tell when a person approaches him whether they are confident or not. How could they help knowing? The attitude and gestures of a person used to horses or at least unafraid of them are completely different from those of a person who is nervous. The former, on being introduced to a horse, speaks to the animal quietly, walks up to his shoulder, holds out a hand to be sniffed, and then strokes him on the shoulder or neck. All his movements suggest confidence both in himself and in the anticipated reactions of the horse. What does the nervous person do? He approaches warily and if he speaks at all his voice does not sound relaxed. He stops quite a little distance away and holds out a hand. If the horse pushes his nose forward he jerks the hand back and takes a step backward usually asking "Will he hurt me? Does he bite?" The usual reaction on the part of the horse is a snort of derision and a departure to other points.

While mounted the rider communicates with his horse most exactly through the language of the aids. The horse has had long training in learning just what is meant and what he is expected to do when the rider uses his reins, legs, back, and the distribution of his weight in specific ways. A highly sensitive horse under a skillful rider generally learns to anticipate what the rider wants him to do before the rider has realized that he has given any signal. To combat this the rider may have to use definite restraining aids to prevent the horse's acting too quickly.

Mr. Butterup was particularly good at this. I usually rode him in our exhibition Dressage Quadrille Team. There were many movements in this in which the eight horses had to change direction or change gait simultaneously. One of these was when the horses, being in line coming down the center at a canter, had to break out simultaneously the first horse to the right, the second to the left, etc., all making a half-circle back to the long sides of the ring. Of course Mr. Buttercup knew the ride. He also knew that he wasn't supposed to turn until told to do so because I had to wait until all were in position. If I remembered to keep my weight exactly even he would do so but, if in glancing over my shoulder to be sure the last horse was where he belonged I unknowingly shifted my weight ever so slightly, the little stallion would bear to the right, and canter the turn.

Since this language is so clear to the horse, how can he possibly not know the difference between an experienced rider and a beginner? All the latter knows or thinks he knows is that you kick a horse to to make him go, you pull as hard as you can to make him stop, and you squeeze your legs all the time to keep from falling off, usually hanging on to the pommel with both hands as well, so that the reins are slack and the horse has had that valuable line of communication cut! Could any horse, no matter how dumb, possibly not know that this rider doesn't know what he is doing, has no control, and that it is the horse that is in command? The flapping of the legs and the hard pull on the sensitive bars not only disturb and confuse the highly trained horse, they are torture to him. It is for this reason that the instructor gives an inexperienced pupil a thick skinned, insensitive mount that will not be disturbed by this punishment. If, by chance, the neophyte finds himself on a highly trained animal he won't be on him long!

Yet again and again one hears, "I don't know why it is, but every horse I ride seems to know right away that I don't know what I'm doing (or that I'm afraid of him) and doesn't behave a bit the way he does when the instructor gets on him!"

Horses can sometimes read the mind of a rider when the rider is still on the ground. Ariel was one of these psychic horses. She had always been difficult to catch in the pasture for she was not a greedy mare and could take a tidbit or not just as she chose. She was particularly hard to catch when she was off at the end of of a field with a foal at heel. At such times she did not wear a halter because there is always the danger of a foal getting a leg caught in the halter when the mare has her head down to graze.

She was turned out day and night unless it was bad weather and of course I made a point of checking her at least once a day. I did not check her myself at meal time—she came to the stable each morning and evening and had her own feed box in the paddock. But this was a busy time and I had other things on my mind.

Generally when I approached her in the field it was just to make sure she was all right and to pat the foal. But sometimes I might want to bring them in and use the mare in a class or drive her. During World War II, we saved gas whenever we could by using a cart, and Ariel, being a Standardbred with an incredibly fast gait, was one of the favorites.

And Ariel always knew what I had in mind. If I was just going to pat her she would come up as soon as she heard the oats rattling in the pail I carried. If I had work in mind it would take me at least ten minutes to coax her to come.

Perhaps you think I made the mistake at such times of carrying a halter or a shank? Not at all, that would have been a real giveaway. Instead I wore my regular leather belt and once I had my hand on her would slip it off, loop it around her neck, and bring her in that way. I used the same dulcet tones in either case and rattled the oats equally enticingly, but I could never fool Ariel—she always knew!

I can understand this extrasensory perception because I have experienced it myself to a certain extent in reverse. Other trainers have told me that they too have had similar experiences. This ability, to know definitely and without question what the horse is about to do is limited to knowing when a restive horse undergoing handling of which he is afraid will permit the handler to proceed with his task without further resistance.

Let us say that you have to treat a horse that has an infected wound on one hock. The animal is young, not too used to being handled, and very nervous. The whole area around the wound is swollen and sensitive.

Let us suppose further that you are working alone. There is no one to stand at the colt's head, to hold up a foot, or to control him with a twitch. You have assembled your materials and have cross-tied him so that he cannot get away but the ties cannot be too short as he must be able to move enough to give warning before he actually tries to fight the ropes in his fear.

You approach with your pail of hot water and your towels. You let him smell the bucket. You spend at least five minutes stroking the colt on his back and rump. You wring out a towel, let him sniff it, and rub the upper parts of the leg with it. Gradually you are able to get a little nearer to the infected area. Perhaps you can dribble a few drops from above so that they run onto the hock. But you know that you are going to have to bathe the wound itself for at least twenty minutes and then put a wet compress on it. All this time your patient is moving constantly forward and back, then a pause, a swing of the rump away, another pause, and the pattern begins again. Suddenly, if you have developed the extrasensory perception of which I speak you will know and you will know *absolutely* that the horse will now allow you to proceed. And in every case he will.

It is a curious experience, this matter of receiving such a message from the mind of the horse—quite different from judging his intentions from the way he behaves. It is as though some voice in your brain said "now" and you knew at once that the time had come.

Perhaps not many people handle enough horses to have such experiences, but every horseman must learn to communicate with his horse—especially through the language of the aids.

Now let us talk a little more about the way the horse communicates with man. We have spoken of how he does so by the attitude of his ears. He also does so with his voice, nickering when he sees his owner coming; with oats is the most common instance of this, but I once had an amusing and slightly different experience.

I was up in the hayloft of our stable in Connecticut doing the morning feeding. The horses had greeted me as I climbed the ladder and walked over their heads to the far end of the loft to open the loft door. The big grain box was at this end but I always forked down the hay first, walking from opening to opening and speaking to each horse by name as I gave him his ration. Soon there was just the sound of contented munching.

Next I started with the grain. The more impatient animals would then stop munching to paw or to bang on the sides of their stalls imploring me to get on with it. I remember one little pony mare whose name was Shooting Star who used to sit back on her hind legs and beg, reaching up through the hay opening with her velvet nose. Then there was Flat Top. When he came to us he had a habit of kicking the side of his standing stall while waiting for his grain. This was not suitable treatment, either for the stall partition or his legs. I used the well-known method of tying a small hard rubber ball to his leg, swinging it at the end of a foot of hat elastic which was tied just above his hock. Thus, when Flatty kicked out, the ball would bounce back and smack him on the side of the cannon bone.

Flatty had only to try this once and get whacked to understand what had happened. After that he kept his two back feet on the ground but swung his big fat rump against the stall instead!

My system in graining was to fill my pail at the grain box, then carrying a funnel and a measure I would progress down the line of pipes, going back to the source of supply as needed and ending with the pipe at the ladder.

The last horse in the line was Bonny. She was an opinionated but not an impatient horse and always stood quietly until her turn came. I reached what I took to be her pipe, shot down her ration, and then stepped onto the top rung of the ladder preparing to descend. Immediately I was almost startled off my precarious perch by a shrill scream of pure indignation! Bonny, seeing me step onto the ladder, deduced at once that I had forgotten to feed her and waited not an instant to let me know it! Had I not known a word of horse talk I would still have gotten the message!

There should always be someone within earshot of the stable at night, for many things can go wrong. Horses can break out and escape onto a highway. They can undo a latch and let themselves into where the grain is kept where they may overeat and come down with colic or be foundered. They can knock down a too flimsy partition or get loose in some other way and pick a fight with a neighbor. They can get cast in a stall and kill themselves struggling to rise. The various kinds of squeals and snorts or the banging of iron clad hoofs on hard oak will tell the educated listener exactly what is going on and whether or not something had better be done about it.

Now let us go to how the horse communicates with his fellows. The intelligent horse owner will learn this language, that is, he will learn to interpret the communication by the way the horse uses his voice and by his attitudes and gestures. The more subtle ways of communication, which we will look into later, will forever be hidden from us for we have lost this ability which horses like most animals still retain.

Let's see just what we know about the vocal language—how and for what purposes the horse uses his voice to communicate with his companions.

The stallion neighs to alert his fellows to the presence of danger, especially when he senses the approach of an intruder. It is preceded by a period of silence in which the horse listens intently, then comes the shrill whinny that can be heard for a long distance. The Arabs understood this. They preferred to ride a mare for scouting. By watching her ears they could themselves know when she had spotted the presence of another rider, but she was less likely to trumpet her knowledge.

The Arab also had his mare sleep in the same tent with him and depended on her soft nicker to alert him of the approach of a stranger at night.

All horses use their voices to greet a member of the herd or a human whom they know. The sound is a low whicker, though the mare will give a shriller neigh on occasion if she is some distance off. The horse being turned out of his stable to join his stablemates in the pasture will often whinny a loud, cheerful greeting as he gallops toward them and they will answer in kind.

A mare will soothe her newborn foal with soft murmurings and gentle nudges with her nose. When she first brings the foal out to join the rest of the herd she will do so silently as though trying to escape attention. They, however, as soon as they spot the little newcomer will set up a perfect cacophony of neighing which calls all to come, inspect, and pass on the new arrival. The mare will instruct the foal in a low voice as to his behavior but squeal furiously at any erstwhile friend who may try to approach too closely.

The shriek of the stallion as he attacks another who threatens his position as ruler of his herd is piercing and the death cry of a horse in extreme agony is both terrifying and heartrending.

Snorting also is a vocal means of communication and is usually used together with certain postures and gestures when strange horses are first introduced to each other. It is all part of a ritual quite different from the greetings exchanged between friends. A horse never nickers at a stranger once they have gotten within nose touching distance, but he will often squeal.

A stallion, and, occasionally in the spring, a gelding, has a particular way of standing when he scents, hears, sees, or learns through the vibrations of the earth beneath his feet that a strange horse is approaching. The best way to describe this is to say that it is an attitude of complete immobility and absolute attention. His head will be up, muzzle elevated, and nostrils flaring. His ears will be pricked and his weight will be squarely over each leg. Sometimes he will maintain this attitude for several moments. Then he will snort, wave his head, prance a step or two, and take up the same position again. To my mind it is the most beautiful stance that a horse can assume.

In all likelihood the onlooker will not be able to see the object of the animal's interest, but if he knows horse language he will be very positive as to what the horse is saying.

Sometimes one is tempted to think that perhaps, this time, he is mistaken. I am reminded of a fine spring day years ago. One of our stallions was turned out into the paddock adjoining the stable while I mucked out his stall. He had had his preliminary little frolic. He had pawed the ground, lain down to roll luxuriously, stood up, shaken the dust off, and bucked and flung his heels in a capriole. Now he was grazing placidly. All at once, for no apparent reason, he came to attention.

"Look at Meadow Whisk," I said to the youngster who was handling the wheelbarrow, "there must be a strange horse coming." We both listened but could hear neither the sound of hoofbeats on the nearby macadam nor the whinny that a horse gives when it approaches a strange stable.

"Perhaps it's something over in the woods, a dog maybe, that's the way he's looking," said my assistant. "Or maybe it's one of the mares and foals over in the far pasture."

"No," I said, "it's a horse all right but not one of our own. It's one he doesn't know."

We went back to our work and after a full five minutes the stallion relaxed and went on grazing. Some twenty minutes later his nerves were alerted again. This time he took up his position facing up the road, not looking out over the pasture as he had done previously. Still we could neither see nor hear anything unusual. But a few minutes later a horseman came in sight, his mount nickered, and the stallion shrieked an answer. Inquiry disclosed that half an hour before he had indeed approached our property via a little-used trail that ended almost half a mile from our barn. Between our pastures and the end of the trail lay ten acres of woodland with a ground cover of heavy brush. Yet the stallion had sensed that a horse, a strange horse, was approaching. He had taken the particular attitude that told me just what had attracted his attention. By going back to his grazing he had said that the situation had changed and that the intruder no longer presented a threat. Then he had repeated his warning when the rider, having retraced his steps, came by a different route. Had this stallion been running with his herd, every member would have been warned of possible danger and would have been ready to take off if their lord and master deemed it necessary.

When a horse or a herd of horses starts running it is valuable to know why. If heads are up, tails are curled flat back, and if the head circles without trying to go any place in particular they're just running out of high spirits and no one is going to get hurt. If they suddenly scatter, run a little ways, and then come to a halt, something has frightened them. If one horse suddenly starts toward a neighbor with his head outstretched, his ears back and his teeth bared, he means business.

A stallion running with a mixed herd will always carefully round up the mares with their foals, separating them from yearlings, geldings, or colts and direct when, where, and how they may graze or move on. Sometimes, if it is spring and there is no stallion in the herd, a gelding will assume the duties of a stallion, collecting and ordering the movements of his harem.

The gestures and movements that the stallion or gelding uses for this purpose are very reminiscent of those of a border collie. He first circles the herd of mares, running with his nose almost touching the ground if a lady tries to sneak away, acting exactly as though he intended to snap at her heels. If one of the older colts or a gelding has the temerity to try and entice away one of the mares, the stallion will bare his teeth, dash at him, and make a grab for the withers or a foreleg trying to force his enemy to the ground. If this does not succeed he will rise high on his hindlegs and box. All this is accompanied by squeals of rage on the part of the ruler of the herd and shrieks of protest or silence on the part of the intruder. Almost invariably the leader will win—after all, he has *right* on his side, hasn't he?

Stallions can be very clever about which mares they have a right to consider their own and which not. Shoebutton, the ten-hand pony stallion, used to roam in the pastures with geldings and mares of all sizes. There was never any trouble for Shoey had decided long since that anything under 13 hands was rightfully his; anything bigger he had little interest in. He never bothered the larger animals nor did they bother him. One year he was presented with a problem. Up until then the largest of the ponies in his herd was only a little over twelve hands whereas the smallest horse was over 14:2, and then a new mare, an elderly dame just 13 hands 1/2 inch, joined the stable. She was with us for several months and Shoebutton never really did decide whether she should be one of his mares or not. One day he'd include her in, the next day he'd chase her away. Many is the time I have watched with amusement, reading his mind perfectly, as he would start to round her up, nose to the ground, then change his mind and with a flick of his tremendous mane, shy away and go back to the ladies more suitable in size. On another day, perhaps feeling more sure of himself, he would round her up and keep her with the rest. As for Gooch, for such was her unfortunate name, she couldn't have cared less, having lost all interest in the male sex years before, and they in her.

That summer we also had a two-year old bay filly, daughter of Ariel. She was close to 14 hands and still growing, much too big for Shoey, or so we thought. But time proved otherwise. We sold the filly in the fall and the following spring the new owner called to say "Guess what? I went out to the pasture today and Red Error had a lovely little black and white foal lying beside her!"

There is a regular ritual to be observed when two horses meet for the first time. They approach each other with heads outstretched, blowing loudly through dilated nostrils. As they touch noses the blowing changes to sniffing each other's muzzles with deep inhalations, nudging each other on neck and chest and sighing softly as they do so. Suddenly ears go back, heads are raised, a foot is stamped, and both horses squeal or snort loudly. They may break away, stand for a moment, and then return to go through the same routine again. Or they back up to each other and engage in what appears to be a lethal kicking match. Close observation generally shows that they are just far enough apart so there is no actual contact or so close together that all they are doing is bumping rumps. The vocal accompaniment, however, is horrendous. If the belligerents are geldings rather than mares there may be a great deal of rearing and boxing and attempts to bite the opponent's withers.

Whatever the method, rarely is anyone hurt and the fight usually ends with one animal giving in and running, pursued for a very short distance by the other. However, such an experience can be terrifying to the inexperienced owner who is sure that death or a broken leg at the very least can be the only outcome.

Although battles between horses while mounted are not common, loud arguments which may lead to fighting are, and can also frighten the beginner. These usually occur when a group of horses with their riders are standing together, reins loose. Or when one rider comes up to speak to another. One horse after sniffing noses will give the characteristic aggressive squeal. If the riders both act quickly and turn the horses away from each other probably that will end it, but if they panic there may be a mock battle royal in which case there is a good deal of danger—not to the horses but to a rider should he fall off into the melee. So riders should always be told never to stand near a strange horse or allow their mounts to sniff noses. They should watch the ears of their mounts when riding in company and if these go back and the horse tries to snap at his neighbor the rider should quickly bend his horse's head away from his companion at the same time staying close to him. This is done by using what are known as the indirect aids of opposition. If the aggressive horse is on the left of the other, the rider takes his left rein quite short and carries it against the neck, the tension being strong, to the rear and right and directed towards (the rider's) right hip. At the same time he pushes hard against the horse's side with his own left leg. This has the effect of bending the horse's neck to the left but makes him stay where he belongs. If an attacking horse cow-kicks then the rider should use his right rein to the right and a little left rein straight back, together with his right leg which should be carried a little behind the girth to push his horse's haunches away from the horse he is trying to fight with. The horses should be made to continue on beside each other. If it is a biting fight that is in the making, the aggressive horse, with his neck still a little bent, should be ridden a step ahead of his partner until he relaxes and "says he'll be good." If a kicking match, the aggressor, haunches pushed a little away, should be ridden a step behind. A sharp word of admonition can also be given and a slap on the shoulders with the crop if the rider is sufficiently experienced.

Incidentally, all the horses in our stable had to learn to work well in company as we did a great deal of pattern riding. Nearly every new horse had to be trained for this. Those who were fearful were always ridden first in pairs with their stablemates or with horses that had kind dispositions. Aggressive horses that started fights got a different treatment. Such an animal was ridden in a three-abreast arrangement between two other aggressive horses that had been trained to go quietly in formation. Somehow the two outside horses communicated to their new companion that he'd better not start something, or else. It does not work to ride the new animal with only one companion because he figures out that if he attacks he can dodge to the side before his companion can retaliate, but with an animal that will take no nonsense on each side he cannot dodge. Of course, there should be experienced riders on all three horses and the one who rides the new horse should carry a crop and with his spurs keep him up in line at the same time keeping him up to the bit so that he cannot go ahead or back out of position. I have never run across a horse that could not be broken of fighting in this way. Generally fifteen or twenty minutes going around and around at the various gaits and changing direction frequently teaches the new pupil that as long as he behaves himself there is nothing to be feared. However, he should not be ridden by a poor rider until his manners are well confirmed.

Let us now go to the less obvious methods by which a horse

communicates with his companions. I shall not attempt to explain just how they do this but will relate a few things that I have seen with my own eyes, and the reader may draw his own conclusions.

The first concerns our friend Bonny. We had had considerable trouble with horses jumping the low stone walls of our pastures. This was during the time of the Great Depression when nickels were almost the equivalent of five-dollar bills as far as availability went. I simply could not afford to encircle our 18 acres with panel fencing, which was safe but expensive and did not deter a trained hunter or jumper if he thought the neighbor's grass was greener than our own. I distrusted barbed wire fencing and had known many a horse that came to grief with ordinary square mesh fencing, as well as others that loved to lean over the mesh and would gradually bend it so low that they could step over it.

Then in a Western magazine I read about the new electric fencing that seemed to be proving successful. It was cheap, for all one had to do was string one thin wire at shoulder length, attach this to a power source and train the horses to respect it by letting them get a mild shock once or twice. From then on, I was told, my troubles would be over, for the well-known fine memory of the horse and his ability to associate ideas would ensure that he never so much as went near the fence again.

So I bought the necessary materials and we fenced in a field adjoining the stable. I had no idea whether or not it would really work, for ours was the only one in the state. I thought I would start with a minimum outlay and could always fence in the other pastures if this proved successful.

The installation directions suggested that we train each horse separately, first wetting the ground on the inside of the fence at a convenient corner. This would have the effect of increasing the strength of the shock which the trainee would presently receive. It was further suggested that we place a bucket of grain on the outside of the fence just within reach.

Having prepared our theater of operations I led Bonny out to the stable, over to the corner, and pointed out the bucket of feed. I then withdrew to a safe distance. Bonny approached with complete confidence, stretched out her nose for the grain, brushed the wire with her chest, and bounced back as though stung by a bee. Circling with lofty steps and indignation in her eye she tried again, this time very gingerly, with the same result. Furious she dashed away about a hundred feet, turned, and waited to see what would happen next.

As it was apparent that she had learned the lesson we brought out old Easy Going. He went through exactly the same procedure, ending up standing beside Bonny.

Huckleberry Finn was the third pupil. He was a slower moving horse than the other two. He bumbled up and leaned forward placidly to be bounced back on his heels and to take off with more speed and agility than he had exhibited in years.

And now a curious thing happened: the three victims proceeded to hold a conference. There was some snorting and some snuffing. They circled off once together only to come back to their original position about fifty feet from the offending wire. For a moment they stood there motionless. Then Huckleberry moved out slowly and advanced toward the fatal corner. Apparently he had been elected to try once more so that it could be established once and for all if there was really a bee there or whether that new wire with the faint ticking sound had something to do with those most uncomfortable stings.

This time Huck did not bother with the grain. He approached very slowly until he was just within reaching distance. Then he stopped, stood for a minute, made one or two tentative attempts, and finally, stretching his long neck as far as he could, touched the wire with a tender, moist, protruding lip. He got royally stung for his pains and galloped back with the information that this was no bee. The three horses moved off and started grazing quietly. They had accepted the fact that unlike other types of fencing this one was taboo.

Now there is nothing that cannot readily be explained in the above account. Bonny and Easy could both see what happened to Huckleberry and if it was he that was chosen to make a rather uncomfortable experiment it was because by nature he was far below the others in the pecking order. Bonny was definitely an Alpha animal and Easy was far higher up the line than was Huckleberry Finn, a very obvious Omega. Also he had only hit the wire once whereas the other horses had had two goes at it.

Let us proceed. Believe it or not, not one of the remaining twenty-four horses, all of whom were in the stable and out of sight of what had ensued so far, when shown the enticing pail of oats made any attempt to reach for it! As they were led out, one by one, and presented with the wire and the oats, they ignored them both, turned their backs, and joined the first three. Nor, at any future time did any horse knowingly touch the fence! The only exception was Meadow Sweet, daughter of Bonny, who, as related in an earlier chapter, preferred to brave the sting in order not to be left alone in the field, and worked out her own method of accomplishing her purpose with the least possible discomfort.

How was the knowledge communicated to those in the stable? Was it the sound of the galloping hooves which followed after each encounter with the supposed bee? But this does not explain the fact that not only did our stable horses know that the fence was taboo, new horses that had never been exposed also realized it. I was constantly having horses in on trial and one would suppose that these, at least, would be fooled—but not at all. Somehow the home horses communicated the remarkable characteristic of the electric fence and not one ever tried to break through it! Impossible, you say? But it happened.

Bonny is the heroine of another unusual instance of the communication of a specific idea, this one contrary to ordinary equine behavior. She had had one colt and was about to drop another. The first had been weaned for about eight months and had another yearling as a playmate. Neither Bonny nor the mother of the other colt had paid any attention to their offspring once the sad days of weaning were over. The other mare had already had another foal, but Bonny had been rebred somewhat later. Mares, foals, and yearlings spent their days together in the large pasture with whichever of the other animals were not being used.

Unfortunately Bonny's second foal was stillborn. Three days later Bonny became uncomfortable due to a very large flow of milk which was spurting out in all directions. She was standing by herself near the end of the stable and I made a mental note that later I must milk off some of the surplus and rub the bag gently with camphorated oil to relieve her. Then, to my astonishment, the following scene took place. Bonny, who had been rather restless reaching with her head toward her hindquarters, suddenly stood perfectly still, head up, and gazed intently at the other animals playing several hundred feet away. A moment later the two yearlings left the herd and came over to the mare. Bonny stood still looking expectant, with the yearlings, their heads raised in inquiry, watching her. Next the mare turned her forehand slightly away and to my amazement Lark, her first born, approached her hindquarters, buckled at the knees, and proceeded to suck away like a newborn foal! After a moment or two he got up, moved out of the way, and the other yearling did the same! Granted that Bonny had figured out that what she needed to relieve her discomfort was a foal to nurse, how had she communicated the idea to two well-grown yearlings that hadn't known or wanted to know the taste of milk for eight months? And how did she get it across to them that she didn't want to be milked dry, only relieved? Above all, why did she permit a strange yearling to nurse along with her own?

We now come to another incident of animal communication which I do not pretend to explain. My readers will probably think I am ''pulling a long bow.'' I can only say that fortunately I saw it with my own eyes, otherwise I would never have been able to figure out how it occurred.

We have talked before of the propensity of horses and ponies to wander. Some are veritable Houdinis when it comes to escaping from paddocks. They will figure out how to open snaps, slide bolts, maneuver the type of wooden catch common in farm gates which must first be lifted and then slid to one side, etc. They are particularly adept with drawbars which have only to be pushed along until one end drops to the ground.

Of all the animals which we have owned, Shoebutton was the cleverest at figuring out such things. Tethered with an ordinary tether chain snapped to the low ring on his halter, he quickly learned to put his head to the ground and open the snap with one foot. When we put a tight strap around his neck with a ring on top to which the tether was snapped, he learned to pull the tether pin up with his teeth and leave, pulling a twenty-foot chain behind him. It was impossible to keep him in an ordinary stall. We had to build him a special one with a one-piece door which opened inward and was fastened on the outside.

We go back once more to the Depression days of the thirties several years before my introduction to the electric fence. Our horses, of which there were then no more than eight or so, were housed in the old barn which we found on the place when we bought it in 1923. Adjoining was a small paddock where the seven or eight ponies, of which Shoebutton was one, were kept. The fencing around the paddock was sheephurdling and the ponies could neither jump it nor roll under it. Unfortunately the entrance had only two drawbars which slid into iron brackets on the posts. The ponies had long since learned how to take these down, but as they usually only escaped into an adjoining pasture, the result was not too serious.

Then came the time when they took to wandering further and threatening my neighbor's lawn and swimming pool, as previously described. I knew that a proper gate with a chain and padlock was the only good solution but this would take time to install and meanwhile I decided to see if I couldn't "fox" them for a few nights with the materials at hand.

The drawbars were two-by-fours. I decided to doctor them by driving two heavy spikes an inch and a half apart in each end of each bar. These were placed so that when in place the spikes dropped down on either side of the iron brackets. In order to slide a bar along it was necessary to lift both ends at once so that the three inches of spike which protruded would pass through the brackets.

Having tested them and found that it took quite a bit of maneuvering on my part to get the bars off once they were in, I "told" the ponies there was no use in their trying to escape as that was no longer possible.

But I reckoned without Shoebutton and his authority as ruler of the herd. I was dressing next morning when I chanced to glance out of my bedroom window from which I had a good view of the paddock. I noticed at once that one end of the top drawbar was no longer in place, having dropped to the ground, and that there was a curious formation of ponies around the barway, all staring at it intently. Directly in front of the remaining bar stood two ponies side by side and about eight feet apart. Behind them was Shoey, waving his magnificent mane which hung almost to his knees, in a commanding fashion, and behind him were gathered the rest of the herd, obviously interested bystanders. As I watched, the first two ponies pushed their heads under the lower bar and, lifting it they gave a simultaneous movement toward the left which slid the bar the necessary few inches, the end dropped beside the end of the upper bar, and immediately Shoey herded the whole bunch out!

Bonny is the heroine of a number of incidents in this book. She could be counted on to win at any game or whenever she was shown. She could express herself very clearly both to humans and to horses. She is shown here at the age of four. She worked for the school into her late twenties, was then retired, and lived on well into her thirties, still able to do light work in spite of having been severly injured at the age of fourteen.

I am not prepared to swear that the ponies acted on command from their leader but the whole scene indicated that this was what they were doing!

The final instance of cooperative behavior which must have included organization and communication of some kind has for its protagonists Bonny (of course), Sky Rocket, and Lark as well as a supporting cast of six or seven others, all females.

The *mise en scene* is some ten or twelve acres adjoining the stable and the time the first really warm day in May. As I have mentioned, whereas generally only stallions bother to assemble a group of mares and keep them separated from geldings, occasionally a gelding will assume such a role, especially in the early spring. This time it was Sky Rocket, then eight or ten years old.

Our animals were not permitted to run in the pasture at this time of year until the mud had dried and the footing was good. Failure to adhere to this practice meant not only the loss of much grass owing to the soft ground being damaged by galloping hoofs, it also ruined the takeoff and landing areas of our outside hunt course which was laid out in two of the larger fields.

Naturally the horses couldn't wait to have the day come when they could gallop at will again, getting the winter kinks out, rolling in the soft spots, and indulging in that delicious new grass. On the day of which I speak conditions were finally right; it was a Sunday when no regular lessons were given and I decided to put at least some of my hard-working animals out.

I started by putting out six mares, including Bonny. I then put Rocket out. Rocket, tail up and eyes flashing, was so affected by the time of year and by that lovely harem that he forthwith decided that that operation he had had six years before was for the birds and proceeded to round up his ladies. Meanwhile I had gone back to get Lark, now a three-year-old.

Now I must explain that the previous year Lark and Rocket had been the dearest of friends, real buddies, and any time that they were out in the pasture together you may be sure that they would either be playing games with each other or grazing side by side. It was for this reason that I had decided that Lark would be the ideal animal to turn out with Rocket and the gals. I planned to let this little bunch have their fun for a half-hour or so and then bring them in and give some others a chance. Green grass can cause collywobbles if too much is eaten before the horses are accustomed to it. But to my amazement, the instant Rocket saw Lark his ears went back and with bared teeth he darted after him! Not only did he not allow him to approach the mares, he chased him out of the field, through a gap into the next field, and then ran him five times around the outside of the 300-foot outdoor riding ring!

Poor Lark, he could not understand to save him what it was all about. Rocket, his buddy, to come at him as though he wanted to chew him up into pieces. Being longer legged than the older gelding, Lark was in no danger of being caught, but he ran and ran fast, only throwing a despairing glance over his shoulder now and again, hoping that it was just in fun. But Rocket's expression left no chance of doubt and at last the colt took refuge in a six-acre lot of woods and underbrush which ran behind the pasture.

Rocket then came back to his mares, moved them along slowly nose to ground in true stallion fashion, and finally held them in a corner where the grass was good. Nothing more happened for several moments. Suddenly Rocket came to attention looking up the length of the field. Beyond a stone wall some three acres away was another two-acre lot which sloped gently upward and was known as "the peach orchard." There stood Lark facing Rocket and equally at attention.

No vocal communication was audible but the flaring nostrils and their positions told me that some silent conversation was going on. Nor was it difficult to deduce what that was. Lark was pleading, "If I stay here can't I come out of the woods and have a little grass, too?" and Rocket was saying, "By no means, I'm the boss here, you get out of my sight! And mind you, stay there, too!"

After several minutes Lark turned and slunk back into the woods and Rocket went back to his mares.

Now I always believe in talking to my horses whether I expect them to understand me or not and this time I couldn't help admonishing Bonny for allowing her son to be so mistreated.

"How can you let Rocket bully Lark that way?" I said. "After all, the poor little fellow has as much right to grass as anyone else and there's plenty for all!" Naturally I got no answer nor any indication that I had been heard. Nor did I expect it. However I continued to stand leaning against the fence in the warm sunshine and watch the animals.

It must have been about five minutes later that I noticed that Bonny had moved a little away from the other mares. And as I watched she increased the distance by ten or more feet. She didn't walk away in an obvious fashion, she just moseyed along, snatching a blade of grass here and there. Rocket was on the opposite side of the herd and did not see her. Fascinated, I watched some more. Little by little Bonny separated herself from the others and neared the patch of woods. Finally came the moment when she disappeared into it.

"Well," thought I, "she's taken my advice and gone to see what's happened to Lark. How nice!"

Again there was a period of inaction, but wild horses couldn't have dragged me away from my post of observation. What would be the outcome? Was Bonny so altruistic that she would forego the fresh grass to stay with her lonely son? Most unlikely! Perhaps they would come back as a team and challenge Rocket?

But what did occur was so entertaining and so unexpected that it has been etched deep into my brain and I can see the whole play as clearly as though it had been enacted yesterday instead of 38 years ago!

What occurred was this. After perhaps ten minutes Bonny, who had sneaked away so quietly, suddenly made the dramatic entrance of a true prima donna. With her tail flat along her croup, the long chestnut hairs blowing like a plume, she dashed out of the woods and straight for the grazing herd. Shrill screams accompanied her coming and fire flashed from her eyes. "Everybody come!" she was saying, "you'll never guess what I saw in those woods, never!"

Of course every head, including Rocket's, came up and as Bonny ploughed to a halt on her hindquarters, they circled around her excitedly, prancing and whinnying and giving her all their attention.

At this point I chanced to look beyond them. There, sneaking quietly along the edge of the woods and heading for the peace, quiet, and security of the stable, was Lark. Bonny's diversionary maneuver had been completely successful!

VII

MENTAL ATTRIBUTES OF THE HORSE

We now come to a subject which is of the greatest importance and interest to every horseman, but especially to those who wish to advance the training of their mounts.

Exactly how does a horse's mind work? What are his abilities and what are his limitations? How does he learn? In training, what methods and situations are to be avoided? How is resistance best overcome? What are the qualifications of a good trainer? What are the various methods that have been used all over the world and what is the basic approach in these and why? Before answering any of the above we must study the aptitudes and the natural reactions of the horse to training. Thus we will avoid mistakes and can hope for success.

In the past there have been authorities who placed the horse's mentality very low in the scale. They judged by the size of the brain in relation to the size of the animal, by his tendency to become hysterical in times of stress, and possibly because he showed little aptitude in solving the type of test generally used in laboratory experiments which involves solving a problem to get at some food.

Yet it is not the size of the brain alone that counts, the convolutions

155

characteristic of a more developed type of mentality are also important and the horse has his fair share of these. As for his hysterical reaction when prevented from fleeing in the presence of fear and of fighting restraint, this is a basic behavior pattern which enabled him to survive as a species when so many other animals became extinct. And does not man also, when the stress becomes unbearable, tend to get hysterical or to go mad?

As for tests which demand that the horse solve some problem which will enable him to find food, these may fail because in primeval days his natural food, foliage or grass, was either plentiful or required that the herd migrate slowly following a continuously diminishing supply.

On the other hand, every horseman knows that horses, and especially ponies, can be diabolically clever in learning how to get into the grain room or to break out of a paddock in search of greener fields.

And consider, too, the ponies of the far north that must subsist through the long cold season by digging through deep snow for buried roots. This is instinctive, · of course, but it must require certain intelligence as opposed to stupidity to know where to dig.

It has also been offered as a sign of the horse's stupidity that he will sometimes stuff himself with grain, if permitted to do so, to the point where a fatal attack of colic can be the result, or a less fatal but just as serious ailment, founder, can render him lame forever.

Here it should be remembered that grain, except in small amounts at harvest time, is not the horse's natural food. Also that because of his small stomach and the necessity to keep continually on the move, the horse has been conditioned to graze almost continuously. How then could it be expected that in the few short centuries since he has become the servant of man, living in a way totally contrary to that which his previous 50,000,000 years had conditioned him, he should change these basic behavior patterns? What is remarkable is that he has been able to adjust at all to being kept confined and fed on concentrated food, to carrying weight on his back, not only on the flat but also over high obstacles, and to running on hard surfaces, the natural contact of his foot to the ground being interfered with by shoes which prevent his frogs from acting as they were intended so that the bones of his legs and their tendons have had to absorb shocks and strains for which they were never designed. That he is better developed physically and mentally as a result of these demands shows a most remarkable ability to survive.

Of late years, scientists have become much more interested in studying an animal's behavior in his natural habitat as opposed to his behavior under laboratory conditions or in zoos. This has meant expeditions to the parts of the globe where such animals live. Scientists have then spent months or years living in these remote parts and have come up with fascinating details never imagined before.

Paradoxical though it may seem, the horse does not lend himself to scientific study of this sort. To begin with, he no longer roams the earth living as nature intended him to do. And to keep a sufficient number of horses for a long enough time for this purpose alone would not be feasible. Secondly it would require that the scientist also be an expert horseman, for the natural habitat of the horse is no longer the jungle but the stable. What must be tested is not only his instinctive behavior but his reactions to the modern demands made on him and his ability to adjust to them and to learn. Furthermore, such an experiment would be extremely expensive. Nor can a scientist go to a large breeding establishment and learn what he wants there, for the average trainer cannot train more than a few horses at a time so the majority of the young stock is sold before being trained. If he goes to various training stables, the animals which he studies will already have been conditioned by their early handling so he will have to decide how much this has affected their reactions. And, again, to understand them he must be a horseman himself.

No, the person who really comes to know the most about the horse and his capabilities is one who has spent a lifetime with them burrowing through books into knowledge which has been recorded by the masters over the centuries, and, with the realization that every horse is an individual, using this knowledge to learn from his pupils.

Rather than try and compare the type of mentality of the horse to that of other animals, let us examine the things which we have learned about him, both through personal experience and through the experience and knowledge of others, keeping an open mind and with the important thought that there is always more to be learned.

Memory

No one disputes that the horse has an excellent memory. Without it, training would be impossible. In earlier chapters I have cited instances of this ability of the horse to remember things that have happened. I would now like to show another type of ability to memorize or remember distinctive patterns of movements and use this memory to their own advantage.

For example, there was Flat Top, who has been mentioned before. He was not a particularly appealing animal, having a rather coarse head and thick neck and the type of conformation which his name suggests. I think he came from the Pennsylvania Dutch country and was probably a cross between a light draft horse and a Shetland or other small pony.

But Flatty was one of the most sagacious animals, one of the most useful, and one of the most endearing that we have ever known. He was perfectly voice trained on the lunge and was used to teach beginners to canter. The routine was to put the rider, probably aged eight or so, on his back with Flatty wearing a pad and surcingle and a stirrup leather around his great fat neck. The reins of his snaffle were tied in a knot and were left to dangle loosely in front of the withers; the end of the lunge line was snapped to the near ring of the bridle.

"Now," I would say, "I want you to pretend you're sitting in a swing and you want to make your swing go nice and high. You must sit as straight as you can, but don't be stiff. Remember that to make a swing go high you push with your back and your seat. You already know how to post. This is different and much more fun. Just take hold of the strap around Flatty's neck. When I tell him to, he will canter once around and then he will stop. You don't have to do anything at all but relax and see how high you can push your swing."

Everything being ready I would now say, "Okay, Flatty," at which the old boy would, from a dead halt, take up the smooth little canter of the trained circus horse, perfectly cadenced and even. Once around he would go and with no word from me would come to a gentle stop exactly where he started. The delighted pupil would then slide off and it was the next child's turn.

Flatty, considering his conformation, was a remarkably well-balanced animal. He is the only horse or pony I have ever seen that in going down hill extended his stride by pushing his croup under him so that the prints of his back feet were further ahead of those of his front feet than they were when he traveled on level ground. And I once saw him, when loose in the pasture shepherding a group of mares, rise on his hindlegs and move *backward* toward an interfering gelding!

Flatty had the usual good memory. Some time previous to his joining our band he had been mistreated around the head and he was very headshy, especially with men or boys. He was not afraid of other things, however, and loved parades where we sometimes painted him up like a zebra.

In class he took good care of himself, always traveling at the same rate as the horse ahead but a good six or eight feet behind instead of the prescribed four feet from nose to tail. He was not afraid of formation riding and was popular especially as a member of the Musical Ride Team. And this brings me to how Flat Top used his good memory to save him steps.

Each year the New Canaan Mounted Troop gives an Equestrian Circus whereby they raise money to support several children through the Foster Parents Plan. There are various exhibitions, some of which take extremely skilled riding and mounts that are highly trained and others, like the Musical Ride, which can be performed by those less advanced in their training. These are memorized patterned rides done to musical accompaniment and are very colorful. The figures vary from simple ones to complex ones and most are executed twice, once to the right and once to the left. Thus in a figure known as "threes by the right and left flank" the column of riders riding single file go up along the long wall of the arena and as each set of three riders crosses the short wall they turn simultaneously down the center of the ring keeping three abreast and well separated so that one horse passes down the exact center, one is halfway between him and the rail to his left, and the other a similar distance to his right. As the next three cross the end they, too, turn and follow in the track of the first three, etc. On reaching the far end, each set of three turns simultaneously, falling in one behind the other in the original order and turning up the opposite long wall to repeat the movement.

It did not take Flatty long to memorize the pattern. As he was one of the smallest on the team he was usually in the last group of three. Having come down the center the first time, unless strongly persuaded by his rider he would then stop at the end and wait quietly for the team to go back to the starting point and come down the second time when, with no urging, he would fall into line to perform the next figure. His attitude was that having done it perfectly the first time he saw no reason to go back and do it again! Nor was Flatty the only pony we had that used his memory to his own advantage. Dandy Beans, a purebred Shetland from Dr. Elliot's farm who grew to the surprising size of almost 13 hands, used to do exactly the same thing!

Bonny applied this form of memorizing patterns to the show ring. She was such a spectacular performer and was so becoming to her rider that she was extremely popular for showing and she enjoyed it no end. It did not take her very long to learn that after having been asked to walk and trot, the class would then be required to canter, first in one direction and then, after reversing, in the other. The second time, after the command "Walk, please," would come the words, "Line up, please." Of course, there was always a suitable interval between the two, giving the riders time to walk quietly and settle themselves before having to turn into the center and form in a line in front of the Ring Steward and Judges.

Having learned the routine, Bonny would obey the various instructions as given and with practically no help from her rider until the command "Walk, please" after the *second* round of cantering was given. On hearing this she would come down from the canter in two strides, walk one step, and then turn immediately to the center, stop, and face the surprised Steward who had not even opened his mouth to give the words, "Line up, please." The first time she did it the young rider, too, was taken by complete surprise and almost bit the dust. After that we were warned and exhibitors were told to be prepared and not let Bonny turn in of her own accord, but if she did, to pretend it was their and not the horse's idea.

We had another pony, Hobby Horse, who was good at learning routines and who loved the game of "Red Rover." We use games a great deal in our school to instill confidence, teach control, and make learning to ride the fun it should be and not the tedious exercise it so often is.

In this game one rider takes a position in the center of the ring while the other riders line up abreast at one end. At the command, "Red Rover! Red Rover! Come over, come over," called by the lone rider in the center, those who are waiting dash down to the opposite end trying to escape being tagged. As fast as a rider is caught he joins the ones in the center and tries to catch those who are still free until finally there is only one rider left who must dodge all the rest if he is to get across safely. Hobby Horse had only to play this game once to learn the rules. As soon as the command "Come over" was heard he would start at a gallop, dodging as skillfully as an open-field football runner carrying the ball and, on reaching the other end, with no signal from his rider would *immediately turn around* and stand stock still waiting to be called back. In fact he switched ends so fast and so unexpectedly that many a young horseman found himself on the ground looking up at Hobby who couldn't understand what he had done wrong!

As might be expected, Bonny also loved games and learned the rules quickly, but she was much more aggressive about the whole idea than was Hobby Horse. So aggressive was she that eventually she was barred from the game of "Musical Stalls," for no matter how poor the rider, whoever rode Bonny invariably won and it was not quite fair to the others.

In this game bars are laid in a row on the ground. They are placed parallel to each other and about four feet apart and as such represent the "stalls." The riders then take the track to music. When the music stops all ride in and try and place themselves in a stall. Since there is one fewer stall than riders, one by one the riders are eliminated. There are certain safety rules. Riders may ride at any gait but, until the music stops, they must not cut in off the track and must all go in the same direction. Once the music has stopped they may turn back or cut in but the stalls must be entered from only one direction and a rider cannot cut across a bar to get a stall. These rules are to prevent collisions in the excitement of the fray.

Bonny understood the rules perfectly and never broke them. When the music stopped, with or without the permission of her rider she would take off for the proper side by the shortest possible path, ducking around the slower riders. As she approached her goal, back would go her ears, her tail would switch or be tucked in tight to her buttocks, and it became her obvious intention to let fly at any member of the equine race who dared try to prevent her from entering her chosen stall. I have even seen her stop if there were more than one empty stall to choose from, stand sideways to them, barring the way, and then step deliberately into the one she wanted while the other riders tried fruitlessly to get their mounts to push past her!

I think the above incidents illustrate some points of the horse's mentality, which is often neither understood nor taken into consideration. They not only illustrate his ability to learn routines, to remember them, and to use this knowledge for his own purposes, they show that the good rider must learn to expect such actions and to prevent the horse's acting before the rider wishes him to, and they also prove that many horses enjoy competition.

Ability to Associate Reward with Obedience and Punishment with Disobedience

Almost as important in training as memory is the horse's ability to associate reward and punishment with obedience and disobedience.

First let us define what, to the horse, constitutes ''reward'' and what ''punishment.'' The reward can be something tangible such as a tidbit, a pat on the withers, or a spoken word of commendation. The punishment can take the form of a sharp oral reproof, a slap, or a cut with a whip. The last two should be reserved for antisocial acts such as a horse's kicking or biting at his companion, charging a person on foot, or a deliberate refusal to obey a command he well understands. This would include the defense of rearing and balking when asked to move out or deliberately trying to unseat a rider. When so employed, the horse must be held on a very short rein with his head held up, and the punishment must follow the offense within two seconds—better still it should be administered while the horse is misbehaving. If the offense is one of balking or of kicking, the punishment should be administered in the form of one or two hard cuts with a flexible bat on the croup. If the offense is charging in the pasture, the person should be armed with a stiff crop. He should stand his ground until the animal is within reach and then catch him across the muzzle, being ever ready to dodge if the horse turns and kicks. The latter rarely happens if the horse has received the punishment recommended, as he will be too surprised to think of further aggressive action.

More important than the above "rewards" and "punishments" are the aids and their use which become associated in the horse's mind with obedience and disobedience. The use of an active aid such as the light pressure of a leg against the horse's side or the slightly increased tension on the bit through one or both reins we will define as "punishment." When the horse obeys, the hand or the leg immediately becomes inactive and the horse receives his "reward."

What is all important is that the trainer uses his aids only just strongly enough to indicate his desires to that particular horse, and that they are relaxed the *instant* he *begins* to obey. If these two rules are not followed, trouble ensues. If the rider uses his aids too strongly the horse will react too strongly and the rider will then have to control this reaction by further strong use of corrective aids. This does not make for a quiet, willing, easily controlled horse. If the reward is not instantaneous, the horse will become confused.

Let us take a few examples. A rider who knows his aids is mounted on a horse that he has not ridden before. The horse is a very sensitive, highly trained animal, whereas this particular rider, though he is not a beginner, has not yet learned to "refine" his aids and use them delicately, according not to what the book and his instructor say, but according to how the horse reacts. The scene is a riding school, the horses being at the walk. The instructor then gives the command to canter. Our rider uses his legs too strongly, so much so that instead of breaking quietly into a slow canter the horse bounds forward. The rider now must use his hands very strongly to restrain him. This is punishment which the horse does not deserve. Had the rider used just a shift of weight and a push with his back together with a light lift of the rein the horse would have gone into the even, slow canter which he had been taught and, when he wanted him to stop, it would only have been necessary for the rider to let his weight settle in the saddle, close his legs, and fix his hands, bringing the horse up into the bit, at which point he would come to a gradual halt.

Just a few misapplications of the aids such as the one related and our erstwhile obedient, calm animal would be jerky, nervous, and inclined to rebel.

Or let us think of the trainer who is teaching his horse to pivot his hindquarters around his forehand. If he applies his aids exactly right, asking for one step at a time with a leg aid and stopping the motion with a delicate opposing rein, his horse will move confidently and calmly, lifting and placing each foot as desired. But if he uses his leg aid too strongly the horse will move too fast and have to be restrained with too heavy a rein.

Let us now see what happens with the rider who does not necessarily use his aids too strongly but who does not release them promptly. We have now a beginner on a willing but not too sensitive mount. The rider is at a walk and wishes to come to a halt. He pulls on his reins, the horse responds by slowing down and stopping, but the rider continues to keep tension on the rein instead of relaxing it immediately. The horse, who has obeyed as he was trained is now confused; he has expected to get the reward which he associates with obedience to the rein, namely a relaxation and the ceasing of activity as far as that rein is concerned. If a more sensitive mount, he might respond by moving backward; if a very much more sensitive mount, he might first back up, and if there is still no release, rear. If stubborn, after a few such experiences, he may refuse to stop at all until hauled back.

One thing that many riders do not understand is that the aids are not used with a steady pressure. Rather they are cadenced to the horse's stride. All movement, whether it be forward, backward, or to the side is initiated by the hind legs. If the aids are to act correctly they must be given in cadence, or rhythm, with the movement of these legs. In this way the horse is not pulled to a sudden stop nor plunged into a faster gait. Not the speed of the gait as a whole but each stride is quickened or slowed on demand from the aids, there being a perceptible lightening of them between the demands, which tells the horse that he is being rewarded, and so he continues to obey.

How does the colt learn this language of the aids and to relate the ideas of reward and punishment to obedience and disobedience? His education starts on the lunge. The equine pupil soon learns to move out when the trainer motions to him by raising the tip of his lunging whip at the same time saying "walk," then holding it behind his quarters. He learns to stop on the word of command when the trainer takes a step as though to block his progress holding his leading hand out. Each time he obeys and stands quietly on the track without turning in he receives a word of commendation and perhaps a pat. At the end of the lesson he may be given a tidbit. His only punishment is a sharp word and a light jerk on the

The equine pupil learns his voice commands on the lunge. Later, when mounted, the trainer continues to use these commands before and as he introduces the use of the aids. He thus takes advantage of the horse's ability to associate ideas.

rein if he tries to break away and a touch of the tip of the whip on his quarters or a wave of the lash behind them if he doesn't move out. His lessons are graduated in such a way that new demands are easily understood and obeyed and there is no temptation to resist.

Up to now the colt has been working in a lunging cavesson with stretched side reins. The next lesson is acceptance of the bit. As explained earlier, this is best given in the stall, and only when the colt has gotten used to the feel of the bit on his tongue and no longer mouths it is he allowed to work on the lunge with the side reins attached to the snaffle. These reins should not put any pressure on the bars as long as the colt holds his head in a natural position. If, however, the colt throws his head, there will be an uncomfortable pressure. Almost invariably he will try this at least once. The trainer will do nothing other than to calm with his voice, for he wants his pupil to learn that no matter how he fights he cannot get away from the pressure except by bringing his head back to where it belongs and also to learn that the very *instant* he does so the reward is automatic. This is a lesson far better taught before the colt is mounted when he has many new sensations such as the feel of the rider's legs against his side and, in essence, it repeats the "lesson of the rope" which he learned so long ago.

When he has thus become accustomed to the bit and has learned not to fight it but to give to it and so receive his reward, the trainer, now mounted, has only to accustom him to a mild vibration on it which will tell his horse to slow his gait or to turn (the "punishment" so called), instantly rewarding him by ceasing the vibration when the horse *begins* to obey.

In a like manner, using voice commands which the horse already understands, the trainer teaches his pupil to react smoothly to the pressure of the legs, using them to bring the horse up into the bit, to increase his gait, to turn and change direction, to move his hindquarters to one side or the other or to hold them in position while he moves his forehand, to move backward, etc., always releasing the instant the horse obeys, then repeating the command again and again until the whole movement is executed. Above all, the horse must learn to move out immediately if told to do so for only thus will the rider be able to always control him.

Learning by Rote or Repetition

Once the horse has learned to associate obedience to the aids with reward and punishment and to understand their demands, reacting immediately, the trainer, by means of exercises, imprints these reactions in the horse's mind so strongly that they become instant and automatic. Beginning with the simplest movement such as breaking from a walk to a trot, a trot to a canter, or a walk to a canter and back again, the horse is taught more difficult lessons such as extending and reducing the length of his stride *without* changing his gait. Eventually the horse comes to wait for each command of his rider continuing at a given gait, speed, and direction until told to do otherwise.

The trainer is careful not to keep his colt working too long on one pattern or one movement and to always work as much in one direction as in the other. The only exception to this is when the horse appears better balanced on one hand than on the other, when the trainer will usually put in a little more time developing the weak side until the horse is properly balanced.

There is no set rule as to how long the trainer must spend teaching a horse a particular pattern, movement, or exercise. Very often he will have to invent a special mode of instruction to suit a particular horse, for everything depends on how the horse reacts.

I was faced with exactly such a situation in training Mr. Buttercup, our Arabian stallion. I had bought him untrained as a four-year-old and gotten him for a suspiciously reasonable price. He appeared to be perfectly easy to handle, was well halter broken, and was accustomed to being groomed and shod.

I gave him the usual basic training on the lunge as described. He learned very quickly, shifted easily from the lunging cavesson to the hackamore with the snaffle under it, and from the bitting harness to the saddle. The time had now come to mount him.

Generally when a horse has developed this sort of confidence and has become accustomed to the weight of the saddle and to the feel of the rider leaning on it from the ground, mounting causes very little disturbance, although the horse may step forward quickly to steady himself and balance against the unaccustomed weight on his back. But Mr. Buttercup was different. The instant I gathered the reins and lifted my left foot a few inches off the ground, his eyes rolled in terror, his ears went back, and he tried first to rear and then to plunge forward.

I quieted him, tried once more, and got the same reaction. Obviously at some time in the past someone had tried to get on him without warning and had frightened him. I now understood why I had been able to buy him so cheaply. His former owners, who were not too experienced in handling highly bred animals, evidently had considered him an outlaw.

I made no further attempt to mount him that day, but soothed him down, led him around until he was quiet, then put him back on the lunge for a few minutes before putting him away.

The next day I came out armed with a large handful of cut-up carrots which I stuffed into the pockets of my riding jacket where I could get at them easily. After fifteen minutes on the lunge I took the lunge line off, leaving the hackamore and snaffle on, the reins up over his neck. I then started walking around the hall with the little stallion beside me. At the end of every four steps I stopped, gathered the reins as though about to mount, handed Mr. Buttercup a bit of carrot, lifted my toe two inches off the ground, restraining any forward movement with the reins, and quickly handed him another bit of carrot when he stood still.

The first time he tried to break away. The second time he only looked as though he might be going to try to do so, but when he found that to get the carrot all he had to do was to stand still and that a lifted toe did not necessarily presage a frightening experience, he caught on to the game. Now, the instant I stopped, he did too, turning his head inquiringly for the expected tidbit then, waiting a second for me to lift my toe, he would turn again for the second one. I worked about twenty minutes with him going in different directions and crossing the arena or making little circles, always stopping every four steps and lifting my toe a little higher and a little higher each time, until by the end of the lesson I could raise it to the height of the stirrup iron and he did not care.

The next day, after a little lunging, we repeated the lesson, adding the step of my putting the point of my toe in the iron without resting any weight on it. Butter was now getting three bits of carrot each time, one when I stopped, one when I raised my foot, and one when, taking the iron in my hand, I inserted my foot in it.

On the third day, I put weight on the stirrup tread without actually raising myself off the ground, holding it there while I handed the stallion his reward. On the fourth day I was able to stand in the stirrup, reach over his back, and hand him his tidbit from the far side. He stood like a rock, even with no one at his head, for now he knew that I was not going to hurt him and that by standing he would get this tidbit.

On the fifth day, I finally swung my foot over his rump and settled lightly down in the saddle. I had stopped on the way up, of course, to give him his reward for standing with me on one side of him, resting my full weight on the one stirrup. Nor did he move when he felt my weight, but waited for the final reward which I handed down to him with many words of praise.

The above somewhat detailed description is what I mean by saying that there are no set rules when it comes to the specific training methods required by a specific horse. They must be tailored to fit the individual, the whole purpose being to teach the horse to have confidence in his master and to obey without fear.

Mr. Buttercup never required any other unusual handling and later became one of our most prized dressage horses, as well as learning to jump well enough and confidently enough to be a member of our Formation Jumping Team. In this ride, eight riders mounted bareback execute an intricate patterned ride over obstacles composed of a thousand inflated balloons. It is performed to music and the riders never break the canter from the time they enter the arena, riding in pairs, fours, eights, coming from opposite directions over the same obstacles, following a serpentine with turns less than ten feet in diameter, criss-crossing with the nose of the horse of the following rider passing barely a foot behind the rump of the animal that crossed in front of him, and "sandwiching" four riders coming from one direction taking an obstacle at the same time as four others coming from the opposite direction with each horse passing between two horses coming from the opposite direction. And all of this is done to the strains of the "Light Cavalry Overture."

The same approach I used with Mr. Buttercup, i.e., assuming that any sign of unwillingness on the part of the horse is an indication that the previous preparatory training has been insufficient or incorrect and that the lesson must be simplified, is used in teaching a young horse to jump. Every new step must appear easy to him and it must be presented to him in such a way that it is more difficult for him to disobey than to obey. Calmness on the part of the horse as well as the rider is of the utmost importance. The horse first learns to walk up to a low bar and hop over it, coming to a walk on the other side. Then he learns to trot around and in between a number of such low bars taking them or avoiding them, stopping, backing up, and then taking them, etc., as the trainer dictates. Not until he is perfectly calm, trying neither to increase his gait and rush the obstacles nor to avoid them, should he be allowed to approach at a canter.

Gradually the obstacles are raised only a few inches at a time and never until the horse takes the previous height calmly and willingly. Variety is introduced by using cavalletti, changing the position and appearance of the obstacles, changing the location of the schooling area, teaching the horse to jump in pairs, tandem, and three or more abreast, etc.

If at any time the trainer meets with determined resistance he must check his methods for the fault. He has either asked his pupil to do something for which he is not prepared or he has persisted too long on one lesson and the horse has become "bored" or "stale," for, like a person, the horse can become resentful and balk if he is asked to take the same jump over and over again. Also many riders do not realize that even a low jump puts strain on the horse's delicate legs, legs which were never intended by nature to receive the jolt of landing with the weight of the rider.

This problem of going sour or stale should also be taken into consideration in working on the flat. Variety can be introduced by changing the order and location in the arena of the exercises, by riding in company doing drill figures, by trail riding alone and with companionship, and so forth.

To go back to the horse that is learning to jump. Many people start their colts over obstacles while their bones are still soft and have them jumping in shows or hunting at the age of three. Because the horse does not go lame they assume that no harm is being done. But they forget that the bones do not reach their full growth and hardness until the age of five.

Breeders who raise hunters and jumpers feel that four years old is the earliest age at which a horse should carry weight over even the lowest hurdle and that his serious training and showing should not begin until the age of five. Even the four-year-olds begin by free schooling with no rider, either in a jumping lane or a Hitchcock pen.

The latter is a circular or oval shaped pen with high fairly solid fencing. The trainer stands in the center and the horse is trained as though he were on a lunge to circle on the outside, taking whatever obstacles are before him. It is also possible to train some horses to free jump in a closed arena or hall. One of our best horses, a Thoroughbred that had been ruined for jumping before we got him by being forced to jump four feet and four feet six in a jumping lane when he had never seen a jump before, learned to take a series of jumps on command, stop, turn, and go back over them with no excitement, no rider, no lunge, and wearing only a halter.

For the horse that is really terrified of jumping, probably owing to previous bad experience, the best approach is to use cavalletti. These consist of a row of bars placed either on the ground or on standards such as those made by nailing two-by-fours in the form of a cross, or using blocks of heavier material hollowed out to prevent the bar from rolling off. They should be no higher than six inches off the ground. These are laid parallel to each other about four feet from center to center, the exact distance from one to the next depending on the normal trotting stride of the horse. There should be no fewer than eight bars.

The horse is first trained to trot over them quietly. Then an obstacle no higher than eighteen inches is set up in line with the cavalletti bars and eight feet from them. The whole thing should never be against a wall, but in from it so that after taking the obstacle the horse can be turned first one way and then the other.

When the horse learns to jump the eighteen-inch bar from the trot, that bar can be raised three inches and the cavaletti bar nearest the jump removed, taken to the other end, and put in place there. The horse can now put in one cantering stride. Each time the obstacle is raised, another bar is changed from the front to the approach end of the combination.

And the rider is careful to insist that the horse wait to be told which way he is to go after jumping and that he maintain an even trot all the way over the bars.

Let me again emphasize that boredom is to be avoided and that horses must not be overschooled. Many trainers believe that horses should not be asked to jump higher than three feet more than three times a week and never be asked to take more than twenty such obstacles at one session. Also that on the days when the horse is not ring jumping he should, if possible, be ridden cross-country up and down hills at the walk and trot to condition his muscles, over low obstacles, broad obstacles such as ditches, unusual but natural obstacles, fallen trees, etc. He should be taught to wade streams and pick his own path through rough trails.

Confidence

The horse can learn to develop confidence in his own ability to do things of which he may previously have been afraid. In jumping he will learn it if his program is such a one as described above. However, it should be remembered that before the horse can develop such confidence in himself, he must have developed confidence in his rider to the extent that he obeys the aids without questioning them. This depends first on whether the lessons of moving out in response to the legs and accepting the bit have been established before there is any attempt to collect the horse or to set his head. Once the first is thoroughly ingrained so that the horse is under the constant control of the rider and listens for the commands given by the aids, the rest is easy.

Let us take the case of a young horse being ridden along a road or trail who meets with something that frightens him. If he has not received the training recommended he can avoid the bit by backing up, whirling, etc. The more the rider uses his legs the more the horse retreats. In the case of an animal that has been taught always to respond, the rider will have no trouble in pushing him past the object. If his training has included the shoulder-in, it is often useful to employ this movement to keep him moving steadily.

In riding a highly trained horse, the rider must bear in mind that his horse is depending on him to tell him what he is to do. I remember that I was once riding our Saddle stallion Korosko B. I often took him to lead a beginner's class, for his "slow gait" was just fast enough for the trot of the ponies behind me. On this day we were riding along the edge of a road. The shoulder was plenty wide enough and was bordered by a rather high grassy bank. We had just crossed over from another road and I was looking over my shoulder to be sure that all of my flock had gotten across safely. At this point the road made a slight bend to the left and as I was looking over my right shoulder I must have inadvertently stepped a little harder in my right stirrup than in the left. The next minute I found myself on top of the bank looking down at my startled class. Korosko had taken the shift of weight to mean that he was not to turn with the road but go straight on up which he did!

Methods of Training Used by Various Peoples Over the Ages

In the foregoing chapters I have tried to apply what we know about the horse's physical and mental characteristics to the way we handle him. Let us look back and see how this was applied by the Masters in earlier times, keeping in mind that though there have been many different ideas as to the correct method of riding the horse and that we have a wider variety of type of horses, basically he is the same animal he was five thousand years ago.

The earliest record which we have today which relates to a systematic training of the horse are the stone tablets of Kikkulis, a Master of Horse in the stables of a Hittite ruler. They date back to 1400 B.C.

The horses with which Kikkulis was concerned were chariot horses. They were used in warfare and probably also for sport such as racing or hunting.

Not much is said about how these horses were handled before the final selection of those suitable for the use of the king was made, but one judges that they were turned out to pasture.

Kikkulis is not concerned with the breaking and training; he is concerned with the methods of choosing so that only the strongest will be selected, and in their conditioning. The former included severe physical tests of long distances at the gallop (having first been warmed up at the amble) with very little food. Once he has made his selection a most meticulous daily routine is described, its purpose being to condition the horse carefully and slowly. It covers a seven-month period and every step at each gait that the horse takes is detailed, together with the exact amounts of water, hay, and grain he is to receive, exactly when and how he is to be bathed and to be groomed, etc. At intervals the horse was turned out again to pasture for a few days since Kikkulis understood the need for variety and for fresh grass.

The next authority is Xenophon, the Greek. He wrote three books on the horse, his care and training, and on riding. He lived in 400 B.C. and his works are very well known. Over the centuries they have been translated many times and there are still copies of these translations extant. The best known was that of Richard Berenger, Esq., Gentleman of the Horse to His Majesty, whose *History and Art of Horsemanship* was published in London in 1771.

Xenophon's theory was that the horse should be trained by kindness and not by force. He remarks that no ballet dancer would spring beautifully if he were whipped with thorns.

In discussing the early handling of the horse he recommends that the young colt be led through the market place so that he can develop confidence in his trainer and can lose his fear of strange sights. He goes into the selection of the horse in great detail, mentioning the "beauties" and the "blemishes." There is nothing recommended by Xenophon which is not considered right in present-day training and handling of the horse.

The Romans, who succeeded the Greeks, were neither such good horsemen nor so understanding of the horse. They ruled by force with bits which kept the horse's mouths open by means of rough wheels which acted on the bars, for example. Horsemanship as an art declined under their influence.

Federico Grisone was the author of the next book of importance on horsemanship and training, which was written in 1569. Unfortunately, it recommends brutal methods rather than the type of training that Xenophon stood for. There were other Italians, among them Pignatelli and La Broue (La Broue also produced a book), who continued to recommend brutal methods, and as a result, one learns by reading La Broue that his horses were always becoming permanently lame or so vicious that they could not be handled at all.

Although he was trained, as was La Broue, by Pignatelli, Antoine de Pluvinel, who wrote a very fine and beautifully illustrated volume on advanced equitation called *L'Instruction du Roi*, agreed with the methods of the Greeks saying that the horse must be treated calmly, his confidence in his master was of prime importance and, especially with excitable horses, the whip should not be used as punishment.

Pluvinel was Master of Horse to Louis XIII and his book is in the form of a dialogue in which the king asks questions and the author answers them. It has magnificent copperplate engravings which show things such as the first use of the "pillars" in training the *Haute Ecole* horse. It was he who introduced the use of pillars, and we see how the horse, fastened with side reins, performs such difficult movements as the *levade* in which he balances immobile on his hocks, the *piaffe* in which he trots in place, and even the *capriole* in which he first springs into the air and then kicks backward, his hind legs stretched out almost in line with his back!

These and other "above-the-ground" movements were part of the training of horses used in battle, for the life of the medieval rider depended on the training of his horse as well as on his own dexterity as a rider and as a swordsman in hand-to-hand combat.

Since the world felt that La Broue's methods were the correct ones, it is probable that Pluvinel's book would never have been published had it not been for the beautiful illustrations. It finally did come out, however, in 1666, some years after his death.

At the Spanish Riding School of Vienna, Pluvinel's method of training is still carried on, its performers being the world-famous Lipizzan stallions originally bred and trained to pull the coaches and carriages of the Emperors of Austria and to entertain their guests.

Over the next centuries the school of brutality continued to be popular with most trainers. Its methods were exemplified by the Duke of

Newcastle whose school was in Antwerp, he being exiled at that time from England. However, the theory of force was challenged by M. de la Guérinière, whose book *École de Cavalerie*, subtitled *"The Understanding, Instruction, and Care of the Horse,"* still considered the most fundamentally correct work on the subject, was of the school of kindness. He advised his readers to study the disposition, aptitudes, and weaknesses of every horse they rode, and to adapt their training methods to suit the individual animal in order to bring it to the highest point possible in its training. The book was published in 1733 and at this time the art of equitation reached its highest level.

There was a hiatus in the development of fine equitation and dressage during the period of the French Revolution, but then in the mid-nineteeth century, beginning with Baucher and paralleled by d'Aure, followed by Fillis, there was a reawakening of the horseman's art in France. Meanwhile, in Austria, Von Weyrother brought the Spanish Riding School back to the preeminence it had enjoyed in earlier days.

There have been several trainers in the present century who have advocated unusual mehods of working with the horse. One of these was called "The Rarey Method" and was popular in the early 1900s. In this method the horse was thrown and kept helpless while everything was done to frighten him with the idea of proving to him that these would not really hurt him and that he was under the dominance of man and struggling against his master would do no good.

Personally, I prefer the American Indian method, whose intent also was to prove to the horse that man meant him no harm and that he would not be hurt if he did not fight.

In this method a herd of wild horses was first spotted in their grazing grounds. These were animals which had never been in contact with man and so, of course, were very frightened of him.

The tribe then followed these horses, at a distance at first and gradually got the animals used to their appearance. All this time, too, they were unobtrusively herding them toward a previously constructed pen.

The pen was so designed that the horses could neither jump out nor break out. The opening into it was narrow and a long chute shaped like a funnel led to the opening or gap. So wide was the open end of the funnel that the horses could be herded between its walls before they were aware of the trap.

Once inside the pen, animals not wanted were escorted out. Then came the work of the specially trained horse breakers. These men knew the location of certain nerves on the nasal bone of the horse and on the crest. Working with a short, thin rope, the horsebreaker adjusted this so that it bore directly on one of these nerves. A little jerk told the horse to stand still, for it was painful, the pressure being instantly released when he did so. The Indian trainer holding one end of the rope in his hand, ready to tighten if need be, now gained the horse's confidence by stroking him and talking to him and patting him. This continued until the horse would stand quietly not only for this but for being rubbed with a blanket or having one flapped around his head. At the end of the day so much confidence had the horse developed that the Indian could mount and ride him away.

It is very probable that the Indian has retained some of the more subtle and less well understood methods of communication among animals which we have lost.

There have also been examples of individuals who could dominate horses by methods of their own which are beyond the power of most people. In 1963 an article appeared in the *Horse and Hound,* a British publication. It concerned a Dr. Hector Geddes, senior lecturer at the Unversity of Sydney in New South Wales. The writer told how Dr. Geddes demonstrated a method of breaking wild horses that had been taught him by an old Australian horseman, Mr. K. Jeffries, who could catch, gentle, and back a wild horse in an hour. Dr. Geddes himself then gave a personal demonstration catching, gentling, bridling, and mounting a two-year-old that had never before been handled. He completed the test in just under an hour, then took off the bridle and remounted the pony to show how quiet it was.

Dr. R.H. Smythe, who mentions this article in his book, *The Mind of the Horse,* does not tell exactly what method Dr. Geddes used, but it was no doubt based on that of the Indians, namely, gaining the horse's confidence until he was convinced that man meant him no harm. Dr. Smythe goes on to say that he doubts if any animals other than the dog and the horse would be willing to submit to such dominance, and to prefer it. This no doubt is true, but it brings up another point. There are many people who have shown the ability to dominate (though not necessarily to develop a preference for domination on the part of their pupils) animals of various kinds. Without this talent there would be no wild animal trainers and no wild animal acts in the circuses.

In my own family my grandmother was such a person. She not only had remarkable control of animals, they were strongly attracted to her. Yet she never seemed to be particularly interested in nor fond of them. I speak now of all kinds of domestic animals and fowl which followed her every time she walked in the ten-acre "yard" of her house in Virginia.

One day when I, as a small child, was walking with her followed by the usual procession of sheep, a mule or so, turkeys, etc., one of the stablemen, his normally black face gray with fear, rushed toward us to tell us that the Jersey bull, well known for his bad temper, had escaped from his pasture and was headed our way.

My grandmother, then aged 82, listened with interest but continued on her way. As always in her right hand she carried a tall shepherd's crook and in her left a dainty little fan made from guinea hen feathers. I trailed along. I had great confidence in the powers of my grandmother.

Presently we heard some warning bellows and then the bull appeared trotting directly toward us, stopping now and again to paw and shake his head, then coming on again. It was obvious that we were on a collision course.

As he approached, the bull slowed down and my grandmother continued calmly on. When they were perhaps fifteen feet apart both stopped and stood still staring at each other. After a moment of silence my grandmother advanced waving her guinea feather fan as one would wave at a fly that had settled where he had no business to do so. Evidently the shepherd's crook was deemed too strong a weapon.

"What are you doing here, sir?" she demanded. "Why aren't you down in the lower field looking out for my shoats the way you're supposed to. Now turn right around and get along back there!"

And the bull did so. We followed him down the road for a quarter of a mile and when my grandmother had shooed him over the drawbars I replaced them and we returned to the yard.

To my mind this is a very clear proof that a person desiring to obtain obedience from either animals, children, or even other persons must be *absolutely and completely certain in his own mind* that he will succeed in doing so; if such is the case, he *will* succeed no matter how fantastic the situation. But he must be as convinced as he is that the sun rises in the east and sets in the west. The moment there is any doubt whatsoever in his own mind, any admission even unconsciously of the possibility of failure then the power is lost and the spell is broken.

I think it is largely this quality which governs the "pecking order" in the animal kingdom. And I think it is the lack of it which makes many parents failures as far as controlling their own offspring goes.

Let us mention now a few authorities who wrote about the science of equitation in the present century. All of these recommend the methods proposed by Xenophon and confirmed by Pluvinel and Guérinière.

In the United States we have a book entitled *Riding and Driving,* the first section by Edward Anderson and the second by Price Collier. It is the first that I have found written by an American that goes into the detailed exercises of the Classical school.

De Souza, who wrote two books and toured the country giving a series of lectures and "clinics," also helped to awaken this country to the importance of demanding something more of the horse than a pleasant ride.

Belle Beach, a well-known equestrienne at the turn of the century, was most interested in the lightness and sensitivity of her horses, saying that she would never buy a horse that she could not hunt using size 60 cotton thread for reins. She was in error on one point for she confidently predicted that the new fad of riding astride which her friends were taking up would disappear since physically a woman would never be able to attain skill in this position.

Piero Santini, a pupil of de Souza, introduced the Italian forward seat in his book *The Forward Impulse,* thus breaking away from the classical idea of the three seats of gravity of the horse, on his forehand, on his center, and on his hindquarters, declaring that only the first was the practical one.

After studying the Italian, the German (Classical), and the French methods of riding, the United States Army, which sent a team of officers to Europe for that purpose, came up with what is known as the *balance seat.* It was most clearly defined by General Harry Chamberlin (then Lt. Colonel) in his book *Training Hunters, Jumpers, and Hacks.* For those interested in the secrets of the "white stallions" (the Lipizzan horses of the Spanish Riding School) their former leader, Colonel Alois Podhajsky has written a very fine text, *Complete Training of Horse and Rider,* in which he emphasizes the importance of slow, careful training in hand and the teaching of the extension of the gaits before any attempt to teach collection is made.

Each year there are new books published on riding but most deal with the rider and few with the horse. Perhaps the event which had the greatest influence on the improvement of the training of open jumpers was the discontinuation of the horse cavalry, which resulted in international competition, including the Olympic games, being opened to civilians.

Up until this point open jumping classes consisted of two or three obstacles, mostly simple post-and-rails, all with wings being set up on each of the long sides of the arena. The horse entered and was ridden twice around. At the most he was asked to change direction once by coming down the center over another obstacle. The open jumper riders brought their mounts in fast and it was very obvious that the majority of the horses were jumping strictly from fear. When the obstacles got too high in the jumpoffs, they were knocked down, and no attempt was made to balance the horse, send him in quietly so that he could judge his takeoff, etc. It was just a slapdash affair with riders getting left behind and coming back on their horse's mouths at every jump.

I will never forget an elimination class I watched one morning at Madison Square Garden. It was some time in the late 1930s. Horse after horse came in, tore around, and either cleared the jumps or knocked them down, sometimes by coming into an abrupt sliding stop which carried them through the whole obstacle to the delight of the audience.

As one horse after another came in, none to clear the course perfectly for by now a few simple turns had been introduced, I noticed a man waiting at the in-gate. He was riding a rather small and what appeared to be very gentle chestnut. I took it for granted that he was merely watching out of interest. When all the other competitors had ridden he entered; his horse took up a gentle, rockingchair canter, circled once, pricked up her ears, and sailed confidently and quietly around the course with no excitement whatsoever. I do not know the rider's name, but I do know that he was a member of the U.S.Cavalry from Fort Riley.

So far we have talked only about methods of training hunters, jumpers, hacks, and dressage horses. We have not mentioned the western school of riding.

As I pointed out in a recent book, *The American Quarter Horse in Pictures*, the whole role of the western-trained horse is different from that of the horse trained for flat-saddle riding, and the training is based on the fact that in working cattle the cowboy needs his hands for his work, roping, bulldogging, etc. Hence the horse must be, and is, trained to work on his own with no help from the rider. The movements that he must know include starting quickly, stopping quickly, turning on his hindquarters, and backing. He must also be able to do flying changes and to balance himself when in the most extraordinary positions. This last is particularly true of cutting horses.

We learned earlier that the horse uses his neck and head to balance. Since balance is of prime importance to the western horse, and since the anticipation of the movement of another animal or animals is the next most important skill which the horse must learn, it is natural that the cowboy rides on a completely slack rein, thus giving the horse the freedom of head and neck that he needs in stopping, starting, and turning.

The rider trains his horse to fear and avoid the bit rather than to depend on it and to interpret the most minimal signals of the shifting of weight, the lifting of his hand or the laying of a rein against the horse's neck, to indicate his desires as to change of direction, slowing, or stopping. These signals are permitted in such competitions as trail horse and stock horse classes, but not in cutting horse classes where the horse must show that he can keep an animal away from the rest of the herd to which he wants to return with no help from the rider.

There are well-trained western horses and badly trained western horses just as is true of flat-saddle horses. The well-trained western horse relaxes at the poll and jaw when asked to stop from a gallop; the poorly trained one does not. Certainly the most well trained and consequently the best balanced is the cutting horse whose mouth is never interfered with. These horses are truly fantastic and show not only physical skill but a very high degree of intelligence; furthermore, they all seem to be enjoying their work.

In the early days, the western riders, having an unlimited number of wild horses which could be had for free and on which they could draw, were rough in their methods. The only contact the yearling had with man was when he was gelded and branded—hardly a pleasant experience. He was then turned out with the herd until the age of three or four when he was brought in, roped, cast, and saddled while on the ground, whereupon the horsebreaker mounted and "rode him to a standstill." This was very similar to the Rarey method, the object being not to gain the horse's confidence but to show him who was boss.

The result was the spoiling of many a horse which could have been turned into a willing and obedient servant instead of an outlaw and injuries sometimes fatal to both the animal and the rider. Those horses that survived of course never again rebelled and became obedient servants.

In the past decade the picture has changed radically. The emphasis in breeding is now placed on producing animals that will win in the show ring or on the race track as well as those which will work on the ranch as stock horses or on dude ranches as quiet mounts suitable for beginners.

For the first purpose expensive blood lines have been developed and the produce, which must have good conformation, tractability, and be without blemishes, have a hundred times the value of the stock horse or dude-ranch animal.

The American Quarter Horse Association, whose registry is the largest of any breed and whose race purses exceed in value those of the Thoroughbred and Standardbred purses, has been one of the influences responsible for doing away with the practicality of the old idea of "breaking" the horse by force. Instead, in its various manuals on care and training, it advises first making friends with the foal and then with systematic and sympathetic handling turning him into an animal that in future may win on the track or in the show ring, or failing the quality for these, will make a good working animal in rodeo contests or on the ranch,

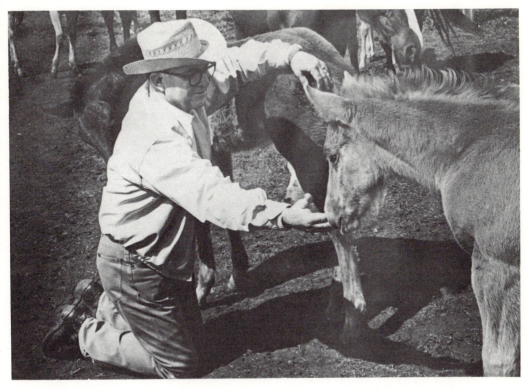

Making friends with the foals. Western ideas of breaking and training have changed radically in recent years largely due to the advice of the American Quarter Horse Association.

or, if not up to the competition in these fields, can still win in trail classes or be valuable on the dude ranch.

The cost of raising a foal has been another factor which has made the breeder handle his colts with care. He simply cannot afford to take the risk of ruining or possibly losing a potentially valuable animal because of roughness. So though the training of the western horse is, and must be, completely different from that of the flat-saddle prospect, the early training of both is similar, being based on kindness. One last comment. Many scientists who have studied animal behavior insist that all animals react and act on instinct, and that to atribute any ability to "think" is an anthropomorphism, an attempt to give them human powers which are beyond them.

Several authorities, especially veterinarians and owners who have had many years of experience with horses and have come to know them as individuals as well as members of a well-classified species, answer this by saying that to agree to this they would have to have a more exact definition of the word "think."

I have just consulted Webster and I find that the definition given there includes the following: *think: to exercise the faculty of judgment, conception, and inference.* Did not Bonny infer, when she heard me step on the ladder, that I had forgotten to feed her? And did she not, from that, judge that this would result in her getting no breakfast? And, finally, did she not conceive the idea of telling me so most unmistakably?

In Webster I did not come across a definition which I think should have been included, that of mentally devising a method of solving a situation and then of carrying it out. a situation, moreover, which included the persuasion of other members of the herd to cooperate.

Such ability I think was clearly shown in Shoebutton's solution of the newly equipped bar problem necessitating team work, with Bonny when she persuaded the yearlings to nurse long after it would have been normal on their part, and in the conduct of Bonny, Easy Going, and Huckleberry Finn in the *Interesting Affair of the Electric Fence!*

VIII

THE EMOTIONS OF THE HORSE

Horses display anger, fear, affection (a mare for her foal), friendship, curiosity, sorrow, dislike, hate, jealousy, and a feeling of happiness, of *joie de vivre*, stimulated by good health and high spirits.

Anger is one of the two strongest emotions displayed by the horse (fear is the other) and is characterized by an aggressive attitude folowed by some form of attack. Horses that become definitely and habitually aggressive toward either humans or their own kind are always those of the naturally dominating type. Those which show their aggressiveness toward humans I am convinced have at some time suffered some traumatic experience at the hands of mankind.

A horse that is naturally aggressive and is abused may become so mean that he will attack to avoid being caught. Such an animal, as the owner approaches, will first lay back his ears and bare his teeth, after which he may charge. Sometimes the charge will end in an attempt at biting, at other times as he reaches his intended victim the animal will suddenly whirl and let fly with his heels.

Curiously enough the large majority of these animals, once you have your hand on them will relax completely and be entirely amenable. This leads me to suppose that they have been broken by the old-fashioned method of being frightened and encouraged to fight until they gave up.

This would account for their submission once they were caught and their extreme unwillingness to be put in a position where the human could apply the "rule of the rope."

I remember an elderly horse of my childhood belonging to my grandmother. His name was Job Trotter, and he roamed loose in the big yard of our Virginia home. My job was to go out and catch him around eleven o'clock, harness him up, and then drive with my grandmother at a sedate and steady jog trot to the village store.

Catching him was the problem. I didn't know the trick of arming oneself with a good heavy stick, letting the horse charge, and before he had time to swing around, to catch him a blow on the nose. This punishment may not make him any easier to come up to, but it will ensure that the next time, on seeing your weapon, he'll be mighty careful not to come close enough to connect with either teeth or heels.

I usually tried to coax him up to me with a handful of grain, taking care to stand beside a convenient tree behind which I could dodge if I saw his ears going back. My grandmother found it hard to believe that any animal so docile in harness could possible have such aggressive tendencies. After being knocked down a couple of times, I got the yard man who did the milking to put Job Trotter in a stall with some corn every morning. This of course, is the only sensible method of handling an animal with such habits, for, as with most of his type, thus confined he lost his vindictiveness.

There are other horses that will show no such signs of active aggression in the pasture but will when in a stall. Such a one was Sky High, previously mentioned. A look in Sky's mouth shortly after we got him showed that at some time someone had put a twitch on his tongue. It must have been of fine hard string or even of wire for there was a deep scar completely encircling the tongue indicating that it had been cut so badly as almost to have been cut in two!

Obviously it was this which accounted for his extreme headshyness. He would allow you to walk up to him in the pasture and to slip the reins of a bridle over his neck. But as soon as you reached for his poll, up would go his head and the average person would not be able to slip the bit in his mouth nor the headstall behind his ears. There is a special techique in bridling such a horse mentioned earlier in this book and Sky High was often used so that I could demonstrate it. Once bridled and saddled Sky was one of the gentlest and most cooperative of creatures, and as long as the bridling and saddling took place out of his stall his aggressiveness was limited to raising his head. However, the traumatic experience which brought on his original objections was evidently associated in his mind with his stall and it was dangerous to go into his stall, especially when carrying tack. Instead, standing outside the box stall door, the top half being opened, one coaxed him to come up close enough for a shank to be clipped to his halter. From then on there was no problem, for when the lower half of the door was opened, Sky would step out and stand placidly cross-tied in the aisle to be groomed and tacked up. One day, forgetting his idiosyncrasy, I entered his stall carrying both a saddle and a bridle. The instant he saw me Sky wheeled and presented his rump. Of course I should have gone out at once; instead I tried to approach his head. Without warning he let fly; fortunately I held the saddle in front of me and his heels hit that instead of my middle. For a moment he had me pinned in a corner while the battery of flying heels continued. Then I was able to edge my way to the door and get out with no damage being done.

The minute I was outside and had put down the tack, Sky came up to me in his usual amiable fashion. I think that his attack was entirely automatic, a conditioned reaction to that dreadful day when he had been so terribly abused.

Some horses, if stabled in a standing stall, though they may not attack violently, will try to avoid being handled by crowding a person who enters it with the purpose of coming up to their heads. This can be prevented by carrying a stick a little longer than the width of one's body. The two ends should be pointed. Holding this stick across the body at waist level one enters the stall so that one point is against the wall of the stall and the other in line with the body of the horse. If he crowds, the horse will feel the point of the stick which, braced squarely aginst the wall, will not slip, and the harder he pushes the harder will be the jab.

There is no question that this anger and aggressive attitude toward human beings is always a combination of natural aggressiveness plus a history of mistreatment.

The one exception is the tendency of ponies to nip brought on by being fed tidbits out of your hand. This is not indicative of a desire to hurt necessarily, but it is a result of the pony coming to expect a tidbit whenever he is approached and his annoyance at not receiving one. It can easily develop into a vice in which the pony will attack. For this reason ponies should not be hand fed. Anything such as a carrot or a bit of apple should be put in their feed boxes and only a gentle pat and a word of praise should be their reward for good behavior.

Stallions nearly always go through a stage of biting as young colts. If they are brought up with other colts this tendency usually wears off, for they take such behavior out on each other. Geldings also like to chew and bite, especially at the age of three when they are teething. Lark at this age used to be restless when I groomed him, so I would put a rubber curry comb in his manger and he would cheerfully chew on this instead of nudging my rear as I cleaned his hind feet.

I have had one pony who was a compulsive biter. His name was Scooter, and I have always suspected that he was a "ridgling" (a male that after being castrated still retained a testicle that never dropped down). He had respect for those who knew his habits and I could stroke his head, bridle him, examine his teeth, etc., and he never even put his ears back. But he would take advantage of strangers, especially children, so when the classes were in session we kept a muzzle on him. He lived to be almost 30 and never got over this vice, which he had when he came to us. To compensate, he was the most ideal beginner's mount both for ring work and for trail work that one could possibly imagine, and for this reason we were willing to put up with his nipping.

Some horses, though as gentle as can be with humans, take out their aggressiveness on their own kind. Generally they dislike being crowded by any horse, be he stranger or friend. Sometimes the dislike only continues until the erstwhile stranger has been accepted in the herd. Occasionally a horse always shows dislike and aggressiveness toward one particular type of animal. Bonny, for example, had an extreme hatred for any and all spotted equines, be they ponies or horses. She demonstrated this again and again, not only by putting back her ears if a pinto approached but, if the place were the riding hall during a lesson under a rider of limited competence, she might dodge out of position and charge the inoffensive animal that might be minding his own business on the other side of the arena.

We all knew this idiosyncrasy of Bonny's and she generally was placed toward the head of the line, with any pinto several horses behind her, but not so far behind that he would be across from her as we went around the hall. I remember one day when my patience was really tried. We had a class working in the outdoor ring when a dealer friend arrived with a horse that he was sure was a perfect beginner's mount. It was a fat western pinto that looked as though he could carry not one but up to three small fry, and I was assured that he would be invaluable especially in teaching the canter, for he had been western trained and would canter in line all day long if required. The dealer said that someone else was looking for just such an animal but that he had stopped by to give me first choice.

So I had the prospect tacked up with a fairly capable rider in the saddle, and they joined my class. Bonny was up toward the front of the line as usual and the new horse fell in a few horses behind. But Bonny was so furious at the advent not only of a strange animal but one of the hated color that she waited until her own rider was a little absent minded and I was paying attention to something going on at the end of the line, then she dodged out of line and *backed up* to within striking distance of the poor, unsuspecting pinto, letting fly with both heels, and almost planting her own rider.

"This," I said, "has got to stop." I halted the class and lined them up in the center out of harm's way. Then I had my assistant mount the newcomer while I got on Bonny armed with a good riding crop. We took the track with the pinto ahead. I rode Bonny up until her nose nudged the fat, black and white rump, her ears were pricked in the most "butter wouldn't melt in my mouth" expression.

Next I passed and rode in front and told my assistant to bump me from behind. Bonny behaved as though she actually enjoyed the experience. We rode in pairs. Bonny couldn't have appeared more friendly. I dropped my reins so that I had no contact with her mouth and we repeated the same maneuvers. But I couldn't fool Bonny. She remembered the crop I was carrying and was not going to be tempted into misbehaving and so receiving the well-merited punishment she knew would be forthcoming.

As a final test I lined the two horses up side by side some little distance from the other riders. Having dropped my reins and stirrups, I then when the standard of horsemanship for young riders was not what it later became.

Gincy had chosen to ride Bonny and we had supposed that this would be a walk–trot class. However, as it turned out, the judge expected a canter as well.

There were many spotted ponies in the ring and at the walk and trot everyone had their party manners on. However, when the Ring Steward called for a canter, the erstwhile disciplined performance turned into a turmoil of confusion. It was instantly obvious that, like ourselves, most of the other contestants had not expected to have to canter and neither they nor their mounts were prepared. As a result, some ponies, their terrified riders clinging to the pommels of their saddles for dear life, charged about the ring at a mad gallop, cannoning into those who were going more slowly. Other ponies balked and refused to canter at all, standing stubbornly in one spot with their ears back. And Bonny? Bonny paid no attention to what was going on, ignoring completely the black and white or brown and white bodies that came at her from every direction. She knew what was expected of her and as soon as the Steward gave the word she picked up her delightful little show canter, weaving her way in and out among the other riders without missing a stride. And when she was pinned with the blue she left the ring with every indication of knowing that she had received only what she deserved! I must add that most horses or ponies that are aggressive toward their own kind are dismounted and, without picking up the reins but letting them hang in loops on her withers, climbed aboard again giving the wretched little mare every opportunity to attack. She just stood there, the picture of good temper.

Of course the minute her original rider got back on her attitude changed completely and there was no further show of friendliness, but this time I stayed alert, crop still in hand, and the instant I saw her ears go back I shouted at her. As a result she controlled herself, though much against her will.

The interesting thing is that Bonny's competitive spirit and her love of winning in the show ring always overcame this dislike for pintos. I remember a class in which my youngest daughter made her debut in an outside show at the age of seven. This was in the early days of showing completely reliable with humans, and vice versa. Bonny and her daughter Meadow Sweet who, though she did not particularly dislike pintos would often put her ears back at the horse that threatened to tread on her heels, were both perfect as far as stable manners went. Sky and others that we have owned that had to be handled with discretion in the stall were the epitome of amiability when it came to their relationships with other horses.

Yet one would suppose that a horse that was actively belligerent and vindictive toward the one would also be aggressive toward the other.

I had long noted that this was not so and it was Bonny that proved that my theory was right. We generally put old Flat Top directly behind her in beginners' classes because we knew that he would keep his distance. He was not spotted, of course, but Bonny, in addition to disliking a pinto no matter how good he was, did not like to have any color horse crowd her in the line. One day, however, Flatty forgot and started to close up his distance. When he saw Bonny tuck in her tail he stopped abruptly and swung his forehand to the center, planting his young rider on the track exactly where, had Bonny kicked as she showed every evidence of doing, she might have hit the child. But Bonny realized this and immediately relaxed her bunched-up rump. In her long life she never ever showed animosity toward a human being on foot. She did feel it fair, however, if a rider who was capable did something to make her uncomfortable, to take revenge.

The most amusing instance of this was something that happened to the same daughter who was described above in her first show.

Gincy was now eighteen or nineteen and I had asked her to pose for some photogaphs which were to be used in one of my books. Among them were those illustrating the rider's positions over jumps and the photographer took a number so that we could show the correct position on the approach, the takeoff, the flight, the landing, and the departure. Of course we had to use the same jump and Bonny was asked to take it a great many times. I could see that she was getting bored.

"Just one last shot," I said, "and in this one, Gincy, I want you to 'get left behind.' " The purpose being, of course, to show what happens when the rider, instead of staying with his horse, comes forward too late and does not stay forward long enough, the result being that as the horse lands the rider's body goes back and his hands fly up.

Gincy performed the maneuver perfectly but Bonny was so infuriated to think that someone whom she knew to be an expert should come down on her back with a jolt and give her a light jab (for Gincy had let the reins slide through her fingers so as not to really hurt her) on the bars that she promptly bucked her off!

Fear

Fear, timidity, and a tendency to hysteria when not permitted the natural response of flight has been discussed in detail, so we will not go into it again here. Suffice it to say that were the horse not a somewhat timorous animal we would not be able to master him, for he is far larger and stronger than we. Or, if we did, as is the case with the elephant, he would never serve us other than as a beast of burden, for it is his excitability to flight resulting from primeval fear and no weapons that makes him trainable for sport, warfare, racing, as a stock horse, and any task which requires speed.

Affection

That horses have any real affection for humans of the sort that dogs display I doubt. That they recognize specific persons and often enjoy their company, particularly if they have no horse companions, and that they sometimes show evidence of liking to be handled and stroked, yes. But that they demonstrate the kind of affection that a mare shows toward her foal, I do not believe. Generally the owner or trainer is associated with tidbits and feeding. The horse soon learns to nicker and come to him when he sees him with a pail of grain or a carrot in his hand. But this is not the same thing. The horse shows no signs of being regretful when his owner leaves him nor anything like the loyal devotion of the dog that never wants to leave his master's side.

We did once have a foal that seemed to prefer human companionship to that of its fellows. We noticed this when the foal was only a few days old. The mare and foal were kept in the front yard rather than in a paddock for the grass was unusually lush, and the mare, not one of ours but a boarder, was not in too good condition. She was a pony mare and we tethered her, as the low stone walls that encircled the yard were innocent of gates. Almost from the day she was born this foal took to leaving her mother to come and make friends with any passing human being. Then we discovered that if one sat down on the grass the foal would lie down and stretch out her neck to be stroked just as a dog does. She would even rest her small head on the knees of her human companion, never moving until forced to do so. But this was the only foal I know of that showed such intense affection for human beings, and as she left with her mother before being weaned I never found out if this affection lasted or was just something related to her foalhood.

The mare's affection for her foal is well demonstrated. When he is tiny and weak she will hardly let him out of her sight. When his teeth come through she never complains as he bites hard at the tender nipple. In the pasture in fly season she encourages him to stand under her tail where she can keep off the pests that his fuzzy little appendage is too short to flick.

Tied up with this affection is the inevitable period of mourning and sorrow when the two must be parted at weaning time. And does this not follow the pattern of human beings in that first must come affection and after that sorrow for the sorrow would not be had there been no affection? And is not the strength of the one dependent on the extent of the other? What we have never loved we do not mourn.

Left to herself in the wild, the mare is possessed, by nature, of the wisdom we as parents lack. Before the next foal comes the mare weans her firstborn. By this time the ties between mother and child are weakened and easily broken. Neither mourns. Is not a great deal of the defiance natural to the middle teens simply nature's own plan of having the young break away from the fold to lead their own lives and thus prevent the weakening of a strain by too much inbreeding?

Curiosity

Curiosity is a trait which is very common in horses and ponies, especially ponies and particularly when they are young. Leave a stable blanket hung over the rail of a paddock fence and, if they can get to it, almost invariably some two- or three-year-old will approach it, nibble at it, and end by taking it off the fence and shaking it vigorously.

Mooney, our little performing pony that learned to count, was particularly mischievous and I well remember the day when a patient Italian mason who was laying down a cement runway was the brunt of his attacks. Mooney started by taking his jacket which was hung on a peg nearby and rushing across the field with it. When this was put out of reach, he came over to see just what this large man, who presented such a tempting fat rump as he knelt to smooth the fresh cement, was up to. After watching for some moments he ventured closer, took a nip, dodged to the side and then straight across the wet and freshly smoothed surface leaving a lovely row of footprints. I did not notice what was going on until it had occurred several times when I came to the rescue, soothed the upset feelings of the mason, and put the pony in an adjoining paddock where he could oversee what was being done but could not interfere.

This bump of curiosity common to most horses and ponies can be used to advantage by the trainer, should he acquire an animal that positively refuses to be caught even when enticed by a pail of grain. The technique is to approach until your prospective prisoner looks as though he were about to run. Instead of carrying a pail of oats or the ordinary carrot or apple you supply yourself with a handful of freshly picked grass. When you have approached as near as you dare, stop, kneel, and without looking directly at the horse you are after, rub the grass between the palms of your hand. Nine times out of ten, and probably the tenth time also, the renegade, fascinated by the rustle made by the grass as you rub, and being unable to overcome his curiosity, will first look in your direction, then drop his head and approach cautiously. Finally he will come all the way up and reach out his nose. Since you are kneeling it will not occur to him to be frightened nor to connect the image of a person in that position with one who wants to catch him. If you move one hand very cautiously, talking to him in a low voice and letting him nibble at the grass, you should be able to quietly take hold of his halter before he knows what you are about.

I remember using this technique to advantage one day when I had been asked by a dealer to stop off and see a likely looking pony. On arrival I found that the pony was in a rather large paddock. The dealer had turned him into it on bringing him back from an auction, not realizing that the little fellow was a devil to catch. Three days had passed and no one had been able either to lay hands on him or to entice him into the stable, but the grass method worked like a charm.

Jealousy

This is an emotion not too common in horses, but I have seen a few instances of it. It is tied in with the protective attitude of the mother toward her foal and, indeed, may be just that and nothing else. Young foals clustered around a human who is petting them will sometimes push each other away, little ears laid back, little tails tucked in, and thrust their own noses forward to be petted, but I am not sure that this can really be called jealousy, rather it is just the desire to get attention. I have never seen it demonstrated between adult animals, one wanting the company of the other for his own exclusive enjoyment, but this may be because it is natural for horses to pair off and, if one of the pair finds a new friend, for the other to pick up another pal.

When one thinks of the emotions with which mankind is saddled which do not burden the equine race, emotions such as avarice, envy, and the desire to hurt or kill just for the pleasure of doing so, one wonders whether in our higher state of evolution we have lost more than we have gained.

Joie de Vivre

There is one emotion that human beings and horses as well can share. This is *joie de vivre* or a true joy of living. Watch a group of colts turned out to pasture for the first time in the early spring and compare them to a school playground. The children make more noise, but the colts play just as gaily. In both instances there will be games and races organized on the spur of the moment. The adventurous boy will show off by balancing on the top of the parallel bars or by jumping from a height; the girl will see how high she can pump her swing, and the colt will spring into the air in a capriole.

But best of all is that marvelous feeling of well being and of communion of spirit when a man rides an animal that he has trained himself to the point where they think as one. This is not a "master-servant" duet. The horse senses the needs of the man and the man responds by letting the horse express himself joyfully. Perhaps it is the hunt field, or perhaps it is the trail, or it may just be the schooling ring, but wherever it happens, both man and horse feel "on top of the world" and each knows that his feeling of *joie de vivre* is both understood and enhanced by the other. This, without doubt, is what the Greeks meant when they imagined the centaur and chose it as a symbol of fine horsemanship.

IX

THE HORSE AS AN INDIVIDUAL AND GENERAL CHARACTERISTICS ASSOCIATED WITH CERTAIN BREEDS

Horses are just as much individuals as are people. In looking back over the hundreds of horses which have passed through my hands, some to remain only the two- or three-week trial period, others which have been my associates for as long as twenty-five years, I cannot remember any two which really resembled each other as individuals. This seems to depend on the exact degree and development of each of the mental and emotional attributes which make up the personality of the individual. And although it is easy to say that all Thoroughbreds are flighty and undependable, all Morgans gentle and reliable, and all Shetlands untrustworthy, there are many, many, many exceptions.

In some ways it might be nice if horses came off an assembly line and one knew that if one bought a certain make (breed) of given vintage (age) one could know definitely what to expect, but how unexciting that would be! And how very limiting: The proper approach is to study the skill and capabilities of the person who is to own and care for a given animal. To what use will the horse be put? Will he be used in company or alone? And what is his ultimate destiny? Is he being bought as a show prospect? As a hunter? As a family friend and companion?

Young people are often carried away by the good looks and apparently high spirits of the prospective purchase. This is fine provided they have sufficient skill to ride and train the lovely creature and are willing to pay the price that a possible show winner will bring. If neither of the foregoing is true, they should look further, remembering that the horse or pony they buy must first be sound and second suitable to their needs in size, training, and personality.

Let us now discuss some general characteristics and some specific equine personalities, perhaps sweeping away some misconceptions and opening the mind to certain possibilities.

The Vicious Horse

I have said earlier that I have never known a horse that was dangerous to handle, excepting one which had been mistreated. However, there are certain horses which, owing to a specific bloodline, probably have a higher *potential* toward turning dangerous under mistreatment than others. Although it is much easier to prevent a horse from developing vicious habits than it is to cure one that has already acquired them, the latter is by no means impossible.

About fifteen years ago I saw a very good example of a cure. I had been told that about twenty miles away there was an interesting breeding farm. The woman who owned it had decided to experiment with "natural" breeding as opposed to the more carefully worked out "controlled" breeding programs of the modern stud farm. In the controlled program a mare is introduced to the stallion at the proper time. She is then checked several times more to be sure that the service was successful and that she is in foal. When her time comes she is kept in a special foaling stall and watched carefully so that someone with experience can be present to assist if necessary when the great moment arrives.

The woman of whom I speak had a large tract of land in Connecticut with a good stable. I do not know just how much experience she had had as a rider but she felt strongly that nature did not intend this type of care and supervision and that she would like to prove that with a band of mares and a stallion she could have equal success by keeping them turned out together at all times and allowing the foals to be born in the open.

Since she was planning to support the operation by selling the weanlings, she decided to look for Thoroughbred mares of good lineage—mares that had had their days on the track and whose legs had given out so that they were unable to go on with a racing career and whose records and the records of their get did not merit their being bought by the top breeders. She also, of course, would need one good stallion, one whose bloodlines and whose prepotence had been proved at least to some extent.

At this time, in addition to the children's school, I ran a school of equitation for prospective riding instructors with an associate. Our program included taking these young people on field trips to see other horse operations. So one fine day in May we set out with detailed directions as to how to find this farm. Following these we found ourselves ascending a hill. At the top the road curved sharply. To our left we saw what were obviously the stables we were seeking. A gate was open and a man stood in the center of the road stopping traffic. Presently around the curve and right down the center of the highway came a band of twenty or more mares with their foals. It was a lovely sight, and we watched as they came toward us, the mares jog trotting and the little foals running along with them. They all turned in at the open gate but the man still held us back. Suddenly, to our surprise, we saw a jeep approaching. It was driven by a young man who held his left hand out of the window and drove with his right. Then we saw that he was holding a short halter shank to which was attached a prancing Thoroughbred stallion! As the jeep slowed to make the turn at the gate, the stallion also slowed down and they passed through the narrow opening abreast.

Naturally we were much amused and also impressed by the obvious good manners of the stallion. We were even more impressed when our hostess told us that she had been advised not to buy this particular stallion because he had had a very bad reputation as a mankiller! In fact, none of the grooms cared to handle him at all, and for this reason he was to be had for a pittance!

I asked her how she managed the supposed miracle and she said that it had been very simple. She discovered that, as is so often the case, the main trouble was that he was being fed full rations and not exercised at all. Furthermore, that he was being used only for breeding, so that every time he was handled it was at a time when he was highly excited. She did not know if he had been trained for the saddle but she did know that he had no confidence in man.

Her first act was to cut down on his grain, and when he had been with her a day or so she turned him into a paddock with an elderly mare whose disposition was good and whose breeding days were past. Running around with the old mare plus being kept on an ample but low-protein diet did much to curb his initial exuberance. Gradually she was able to make friends with him and when he realized that she was not about to go for him with a pitchfork, he calmed down. As he became more tractable she gave him more horses as associates, and spent more time with him herself, leading him around, grooming him, and talking to him. As we could see, that method had been highly successful, for now anyone could handle him with perfect safety.

What About Stallions?

While we are on the subject of stallions I would like to say what I have said before many times—a stallion is not a dangerous animal provided he is trained and handled correctly. We have had stallions of Thoroughbred, Saddlebred, Arabian, Welsh, and mixed bloodlines. Most I have broken and trained myself but some I have not, and little Shoebutton, who came to me first as a three-year-old for breeding, is proof-positive of three things. First, stallions are not by nature dangerous. Second, they can be made so. Third, at least in some cases, a spoiled animal can be rehabilitated to the point where he is thoroughly trustworthy.

Shoey came from the previously mentioned Dr. Elliot of the Belle Meade Pony Farm in Virginia. At the time I had two pony mares which I had gotten from Dr. Elliot, and I asked him to send me a suitable stallion to serve them. When Dr. Elliot shipped us Shoey, he told us that he was quiet to ride and to handle but not to drive him, as he had had a bad experience in a cart and he would kick off any harness.

My children were very young then but they could ride a bit and for the few weeks that Shoey was with us the six-year-old was able to ride him on the trail with me or through the fields by himself while the two-year-old enjoyed her turn on a lead. Shoey was a perfect gentleman and I hated to send him back. In fact, I asked Dr. Elliot if he would sell him, but at that time he was not for sale.

Two years later I got a letter saying that Shoey was now on the market and that he was at a stable in Mt. Kisco not too far from us. We could hardly wait to see him again and, armed with carrots, the children and I piled into the old station wagon and off we went.

The establishment was a really fancy one with magnificent stables, 12×12-foot box stalls, pastures, paddocks, and schooling areas. Everywhere we looked there seemed to be a groom or a stableman polishing brassware, sweeping an already immaculate aisle, or putting a final shine on a glossy coat. "What a fine life Shoey must have had!" I said to the children, "I hope he won't have gotten too used to luxury!"

But when we asked to see Shoey we were taken in tow by a stableman carrying a pitchfork who led us to a little shed a long way from the other buildings.

"You don't want to be letting the children go near the little devil!" he told me, "he's likely to kick them right out of the stall!" I could hardly believe my ears. Shoey? Mean? Perhaps Dr. Elliot had made a mistake and this was not our dear Shoebutton at all. His stable name, as I remembered had been Belle Meade Success. Possibly there was another stallion with a similar name. I was quite prepared to see a totally different animal and to disclaim all knowledge of him; I certainly didn't want a stallion that had to be watched!

When the stableman opened the door I was appalled. It was a tiny shed with the typical shed roof high enough to stand under at the front but sloping back to somewhat less than four feet from floor to rafters at the back. There was no window and no ventilation except when the top half of the door was opened and, since it was tight shut when we arrived, no doubt it usually remained that way.

The stall was two feet deep in muck. Shoey's (for it was Shoebutton) poor feet were so long that they were beginning to curl up at the toes. His magnificent tail which swept the ground and his mane which reached to his knees were a mass of snarls. His body was caked with manure. Here was a pony that had had no care or handling for many a long day. And what a traumatic experience for an animal whose greatest fear is confinement, especially in a dark place, and whose instinct is to be constantly on the move! (Had Dr. Elliot ever known of such mistreatment, his heart would have been broken.)

As soon as the door was open the man started at him with his pitchfork. "Get over there now, before I clobber you one!" he shouted, and the pony put back his ears and retreated to a far corner with his tail tucked in.

Naturally I could hardly wait to get him home. It did not take too long to get his confidence, though he still retained evidences of fear. He was very touchy about his hindquarters, and in mounting he would only permit the rider to stand well forward at the shoulder, facing the rear. He would not permit "monkey drill" nor allow any riding double. And when out in the paddock he was afraid of being cornered, so to catch him one had to hold out a tidbit or offer some oats or a handful of grass—then he would come up willingly. He was not at all headshy, however, so evidently his mistreatment had consisted of being attacked from behind.

Having learned his idiosyncrasies, we were careful to warn his young riders. But Shoey could behave himself even when these rules were broken, if he wanted to. He loved to show off and understood the routine of the show ring very well. I had been away for a few weeks, leaving an assistant in charge, and I got back on what turned out to be a day when there was a nearby show. Several children with various horses and ponies had already gone and I decided to go too and see how they were making out.

I got there just as a class of children's ponies was being judged. The principal qualification of a good child's pony is manners. The class had walked, trotted, and cantered and were now lined up. Shoebutton was right in the middle between two little mares, one of whom behaved as though she might have been in a highly interesting condition, but, true to his training that when you were at a show other things must be forgotten, he was paying no attention.

His rider, I noted, was one of our more scatterbrained seven-year-olds. It was her first show and she was naturally very excited.

One by one each young rider was asked to dismount and mount again. When Susan's turn came she forgot everything she had been taught. She dropped her reins and slid to the ground. Then, standing well back beside his croup and without picking up her reins she proceeded to hoist herself aboard again, giving poor Shoey a hearty belt on his sensitive rump as she swung her leg over. Had it not been the show ring, Shoey would never have put up with such treatment, but now he was wearing his ''show manners'' and stood like a lamb.

For many years Shoebutton and old Jessica, a mare which we also got from Dr. Elliot, were the daily companions of my oldest boy, Skip, and his little sister. When he was eight and she was four they used to go off fishing for the day carrying a pole in one hand, sandwiches and a can of worms in the other. Rubber tire tubes were hung around the ponies' necks for they liked to play in the ponds when they tired of fishing. They both wore bathing suits and little sister also wore a life preserver. They would be gone all day and I never worried for I knew the ponies would keep well to the side of the road of their own accord, something which a bicycle cannot do, and that as the sun got low in the sky I would hear the cantering of the hooves of the ponies and the cheerful voices of the children as they returned having had another happy day.

Shoey was a good little jumper and Skip often rode him in the local drag hunts. If the walls proved too high, he would treat them like Irish banks, jumping to the top and changing legs before he sprang off again. His most amusing characteristic, however, was the way he undertook responsibility in the classes. He was usually put at the end of the line from which he could watch everything that was going on. He very soon learned that certain of the horses and ponies liked to take advantage of the beginners by cutting in to the center, especially if I were busy giving special attention to a rider that was having problems.

If I called a rider to come into the center so that I could check his stirrups, for example, Shoey would do nothing. But, as sometimes happened, should the pony of another rider try to follow the first without permission, Shoey would hastily dive forward with his ears back, whereupon the disobedient animal would return to his place in the line.

Shoebutton was a Shetland stallion who was used as a child's mount for lessons, showing, and hunting from the age of two until the age of thirty-two when he had to be put down because of bad teeth. Toby and the bloodhound lying at Shoey's back feet are about to start off on a hunt.

When Skip outgrew Shoey, younger brother Toby took him over. It was Shoey that was featured with Toby in a movie made by Fox Movietone. This was a "short short" in which they took pictures of the riding school and also showed our bloodhound hunting. Toby was then nine and Shoebutton about fourteen. They were the hit of the movie.

As I have mentioned, Shoey lived into his thirties when he had to be put down because his teeth wore out. Meanwhile, he had carried many, many, children many, many miles, one of the last being Logan, daughter of my oldest daughter who normally rode Jess when Jess and Shoey were the mounts for the fishing expeditions. Except for his propensities in escaping from paddocks and fields, Shoey never gave us any trouble and I still miss him!

Korosko B., The American Saddle stallion, was always very gentle to handle and had a really beautiful disposition. He had one habit, or trait, which was unique. If, while carrying a bridle, one opened the door of his box stall, he would come up and, seeing a bridle, would of his own accord drop his head and open his mouth for the bit even before one held it up!

The interesting thing is that his son, Squirrel, developed exactly the same habit! For many years it has been said that animals cannot inherit learned habits, but recently I have read that this is now being disproven, and certainly Squirrel was an animal who appeared to inherit a very distinctive behavior pattern.

Of all the stallions I have owned and worked with, the dearest and most appealing was Meadow Whisk, the little Thoroughbred of royal lineage who unfortunately went blind from periodic ophthalmia.

He was a four-year-old and completely unbroken when I got him, but had always been handled kindly. My youngest daughter, Gincy, at that time was about twelve and she undertook to train him. He gave her absolutely no trouble and was a particularly apt pupil. He had lovely gaits and great aptitude in jumping. When she had been schooling him over ordinary jumps for only a few weeks she suddenly decided that she would like to show him in our annual circus over a "mental hazard" course. This type of course is designed to show the obedience, training, and courage of the horse. Instead of using the conventional obstacles such as rails, chickencoops, imitation stone walls, metal barrels painted in bright colors, etc., one uses such unusual things as a clothesline complete with blowing clothes, two chairs set facing each other in the center of the ring with their seats touching, a row of stable pails with rocks in them hung on a rail and rattled vigorously as the rider approaches, a stream of water from a hose, a wheelbarrow painted white, etc. Perhaps these things do not seem very challenging, but actually, because they are unusual and because it is much easier to avoid most of them than to jump them, it is far harder to train a horse for "mental hazard" jumping than for ordinary high jumping.

"But," I said when Gincy approached me with the idea, "you've only just started Whiskey jumping and the circus is almost on us!"

"I've got two days," she said. As it turned out she only needed one, for when she put up the course the brave little stallion took every obstacle the first time!

This aptitude for learning, plus confidence in the rider, is one of the most sought after qualities in a horse and both of Whiskey's colts, Meadow Lark and Meadow Sweet, inherited it. Lark expecially never had to be shown anything more than once or twice, whether it was how to take up a gallop from a halt or how to two-track.

Whenever people speak to me of the natural bad temper of all stallions I think of a sight which I saw many times. It is of two little girls sitting side by side on the top of the closed lower half of Whiskey's box stall eating their lunch. He was completely blind by then and they, knowing he enjoyed company, regularly ate their sandwiches and drank their sodas while Whiskey, his head thrust between them, was allowed his share.

Do not feel from the above that I recommend the buying of a stallion for a beginner to handle. Nor even for an intermediate rider to have as his own to care for and to ride without supervision. But I do maintain that the latter can ride a properly trained stallion and an expert rider, even though quite young, can take care of and handle a stallion by himself as easily as he can a gelding or mare.

What About Ponies for Children as Opposed to Horses?

As I mentioned earlier, many people have a prejudice against ponies often because of unfortunate experiences. It is true that many ponies are stubborn, opinionated, bad-tempered, and unmanageable by a beginner. This is also true of a good many potential beginner's horses which may have the additional undesirable characteristic of being nervous.

The well-trained pony of a size proportionate to the rider is a pearl without price. No pony should be bought without being tried out both on his own home grounds and on those of his prospective owner. No beginner should ride either a horse or pony without suitable supervision and instruction. And, whenever possible, when the time comes that the young hopeful can walk, trot, and canter with and without stirrups and reins, can bridle, saddle, groom, and care for a suitable mount; when, therefore, you, the parents, have agreed that he should have one, if at all possible get a neighboring child to go into the investment with you, build a stable for two, and get two ponies for they work ever so much more willingly in pairs than alone.

If your child has been having lessons, have his riding instructor find him a pony. If you, yourself, are experienced and want to teach him (which, generally speaking, does not work) go to a breeder who finishes his ponies and who will give you a choice, agreeing to exchange one that turns out to be unsuitable for another.

Of the pony breeds, the Welsh when they are well trained and settled down are very satisfactory. So, too, are many of the British "moorland" ponies, the Dartmoor and the Exmoor, etc., though these are not common in this country. Shetlands can be very good if they are neither the heavy, British type bred to pull a cart nor the excitable Hackney show type. Very often the most satisfactory is a pony that is a mixture of breeds with perhaps some horse strains as well as pony—it is not necessary to have a purebred.

Of horses suitable for the not too experienced rider, again we find that each individual animal must be considered not so much for his bloodlines as for his own personality and training. The Quarter Horse, or one which is part Quarter Horse, often makes an ideal mount for a fourteen- or fifteen-year-old, being docile, hardy, and intelligent. Morgans, too, are great favorites. For the older rider who just wants something with an easy gait that will carry him along shady trails, the Tennessee Walker will often prove ideal for they slide along at the slow gait, step out at the rack, which does not require that the rider post, and canter all day under "under the shade of a tree."

For the expert rider who wants to try his hand at how far he can carry the education of his horse, there is no limitation. I knew a man in Canada who was a most proficient professional trainer but who no longer worked at it as a profession. He had not a great deal of money to spend and his training area consisted of a rather rough barn and a little paddock not more than sixty feet square. He had kept up his interest for a number of years by going to the auctions when a carload of green horses came in from the west, picking out a couple of likely prospects and, in a year or two turning them into highly trained dressage mounts. The pair that I saw had not the most beautiful conformation in the world, but they were

Sky Rocket is the only horse I have ever known who preferred jumping to anything else. Though exceptionally sensitive by nature and timid from mistreatment, he would take any type of obstacle put before him and was never known to shy out or to refuse.

executing piaffes, the lofty steps of the passage, and would change lead at every step!

Jumping ability is an individual thing and good jumpers are to be found in all breeds, although some breeds are generally less talented, e.g., Morgan, Quarter Horse, and Arabian. Nor is every Thoroughbred a good jumper prospect. Few horses like to jump of their own accord though, if properly ridden, they do not mind taking the hurdles. Open jumpers are apt to lead rough lives being made to go ever higher and to give their all day after day at show after show. The hunter has a much easier life. Manners and way of going being of prime importance, these horses are not rushed in their training and seldom have to go through the ordeals of poling that is the fate of many an open jumper.

I have never had but one horse that I felt would rather jump than do anything else. He was a little fellow of unknown breeding named Sky Rocket. (I have told his full story in a book of that name and here I will only say that jumping seemed to be the very breath of life to him.)

He was about the same size as Whiskey and I bought him for Gincy when it became evident that Whiskey's jumping days were over because of his blindness. My how that horse loved to jump! One day, put into a paddock because of a slight lameness, he spent the afternoon jumping out into the pasture, galloping down to the end, and jumping the electric fence (the only horse ever to do so) landing on the edge of the macadam road, galloping back along that, and jumping back into the paddock again to repeat the performance until we finally cought him!

An admixture of Thoroughbred or Arabian will give sensitivity, but keep the purebreds for those who know what they are doing and prefer this kind of animal. For them, only the most sensitive and most responsive will do.

INDEX